The Better to Eat You With

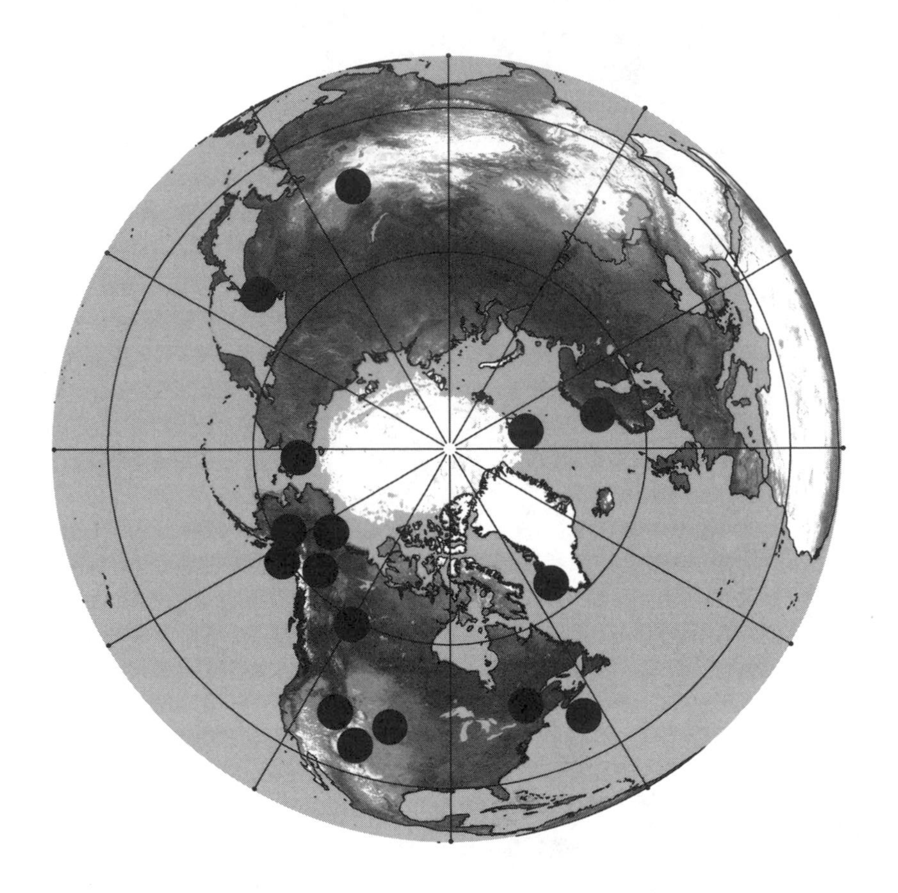

Overview of major northern sites

The Better to Eat You With

FEAR IN THE ANIMAL WORLD

JOEL BERGER

The University of Chicago Press
Chicago and London

JOEL BERGER is a senior scientist with the Wildlife Conservation Society and the coauthor of *Horn of Darkness* (Oxford, 1997). He has worked across the globe, from Africa to Alaska, using quite bizarre but acclaimed techniques to simulate animals and gauge their behavior in the wild. Currently, he serves as the John J. Craighead Chair in Wildlife Biology at the University of Montana.

The University of Chicago Press, Chicago 60637
The University of Chicago Press, Ltd., London

Printed in the United States of America

17 16 15 14 13 12 11 10 09 08 1 2 3 4 5

ISBN-13: 978-0-226-04363-0 (cloth)
ISBN-10: 0-226-04363-0 (cloth)

Library of Congress Cataloging-in-Publication Data

Berger, Joel.
The better to eat you with : fear in the animal world / Joel Berger.
p. cm.
Includes bibliographical references.
ISBN-13: 978-0-226-04363-0 (cloth : alk. paper)
ISBN-10: 0-226-04363-0 (cloth : alk. paper) 1. Fear in animals. I. Title.
QL785.27.B47 2008
591.5—dc22

2008000418

∞ The paper used in this publication meets the minimum requirements of the American National Standard for Information Sciences—Permanence of Paper for Printed Library Materials, ANSI Z39.48–1992.

Through the Jungle very softly flits a shadow and a sigh—
He is fear, O little hunter, he is fear.

RUDYARD KIPLING, *The Jungle Book* (1897)

Contents

Foreword

Modern humans have followed climate and wildlife since time immemorial. Humanity has been shaped and "placed" by climate and the changing habitability of different landscapes. In related fashion, the availability of wild animals and plants for food has figured largely in the human equation throughout history. Climate, specifically the last glacial maximum of twenty thousand years ago, is identified with human habitability, the modern human diaspora, and the gradual settlement of peoples.

Until most recently, climate change has not been anthropogenic; human impacts on land cover were localized and globally unimportant until the nineteenth century. Such is not the case with human effects on wild animals, in two important senses. First, as the cave paintings and glyphs of Africa and Eurasia indicate, humans and wildlife have been painfully but inextricably connected from our first cultural moments. Second, humans may well have been important elements in local—or even entire—extinctions of wildlife, beginning in the late Quaternary.

In any event, humans were both predators and prey, but their impact on wildlife behavior is unquestionably important. One need not accept Thoreau's lament that our contemporary wild nature is but a torn fragment of the original to understand that humanity is a deadly modifier of landscape and wildlife.

AS JOEL BERGER tells us in manifold ways, predation—human and otherwise—shapes the behavior of wild animals. Darwin famously observed that animals on the Galapagos Islands had not yet learned a "salutary dread" of humans, because they had never been taken. Wildlife biologists and amateur naturalists record "naive" primates who apparently have never been attacked by humans or wild birds so innocent that they can be captured by hand. The historical record is full of sad journal entries about wildlife going to slaughter at human hands with no apparent understanding of the predator-prey relationship.

More prosaically, every hunter or fisher or observer of wildlife will tell a tale of the fish who eludes capture because it is too smart, the covey of birds that flies nervously because it has been shot into recently, the turkey that disappears when hunting season begins, or the geese that never come within firing range.

But it is too easy to single out human predation as the only factor in wildlife response. And this volume concentrates on the more complicated tapestry of prey response to fear or threat. I recall a fabulous snapshot of what appeared to be a sophisticated cognitive framework for assessing threat on a trip three years ago to Nagarahole, Karnataka, in southern India. Thirty meters in front of me, a spectacular female Bengal tiger was ambling across an opening in the tropical dry forest—the personification of menace to the spotted deer five meters to the right of me, who stood barking alarms to its compatriots feeding nervously in the grass. The deer focused completely on the tiger, having concluded correctly that I presented no harm. The tiger was nonchalant, perhaps seeing another conservationist in a vehicle and a future meal on the hoof. The responses of both predator and prey suggested elaborate cognition. Tigers have been hunted by humans such as myself, as have spotted deer. In this protected area, in less than a generation, the fear of humans one might have expected from tiger and its prey alike were nowhere in evidence. As a newcomer and nonspecialist, I may have missed an entire range of vocalizations from birds, rodents, and interested others in the little tableau. And what of the possibility that the spotted deer was close to my vehicle because of some discrimination between my threat and that of the tiger, mimicking the Yellowstone elk hanging around the lodge and away from the wolves?

This narrative and its puzzles are hardly confined to India, of course. A continent away, some conservationists working with African savanna elephants will say that elephants know the boundaries of protected areas and avoid hunting grounds. A short course for American bison would be a great service for those individuals who wander off the Yellowstone National Park grounds, converted into "fair game" for hunters and wildlife managers. Bison do not seem to have gotten as far as elephants. Thanks to Berger's experiments, they may be more sensitive to lion calls or mystified by the sudden appearance of tiger dung, but on the evidence, they should pay more attention to wolves and armed Montanans. On the human side, we have examined very little—outside the chapters of this admirable volume—of our own impact on fear and survival among animals we have hunted.

To return to the point, wildlife, it appears, learn from predation. Those who don't are easy prey. Sitting by the Bronx River near my office in the spring, watching as night herons pick off ducklings, it can be quickly concluded that the price of lagging or skylarking next to the riverbank—failure at one's lessons—is death. This dynamic is, in microcosm, what has been called the deadly syncopation of predator and prey. But the lessons are complex, and the learning process, opaque. How does it take place, and how deep does it go? To which signals must a prey animal respond in order to survive? What are the possible impacts of its responses on other animals that live in the same place and may respond to the same cues? What is the gap between the artful dodger and the neurotic that responds to every sound and scent?

More powerfully, this learning, which Joel Berger studies with the enthusiasm of a lifelong naturalist and the methodological rigor of a scientist, is the fundament of a kind of wildlife culture. Berger is addressing a great general question: can the depth and transmission of learned behaviors in nonhuman animals be called culture? More specifically, as I understand him, he wonders if prey are "made" (or socially constructed, as it were) by their relationship to predators. If so, do animals that lose that relationship of fear—one imagines without much grieving—also lose some of their culture and their successful behaviors? In the end, at some deeper level, does prey require a predator? And in reference to humans, does our culture subconsciously require prey?

As a prey animal given voice, I might answer "yes, in general, but please let's not make it too specific." That is, the answer may be good for the group but bad for the individual. In social animals, the individuation of group learning is itself interesting, particularly to a social scientist, of course. What is the learning curve, how fast is a lesson lost, how does survival become culture, in individuals and in populations? What of the differences between social and solitary animals in cultural transmission?

If these questions were not enough to make this volume interesting and rich, the final puzzle is for us as humans. As the "apex predator" or as a member of a landscape community of plants and animals, humanity also wins and loses, learns and fails to learn, in relation to prey animals and the wild circumstances in which they live. This book stands as a prima facie argument for a notion of humans imbedded in wild nature, not apart from it. As a late modern cultural conceit, the separation of the wild from the civilized in human concepts fails humanity and nature alike. The questions Joel Berger poses require humans as part of the equation, as we have been for tens of thousands of years. We humans, too, are shaped by our interactions with wild nature—informed, edified, and warned, as well as fed, clothed and validated. Sadly, from my viewpoint, the growing cognitive and cultural distance between us as predators and wildlife—either as prey or cultural icon—between civilized humans and untamed landscapes, impoverishes us all. Having read this volume, I now realize it may have impoverished the culture of animals, as well.

Steven E. Sanderson
President and CEO
Wildlife Conservation Society

Prologue

An old Russian once told me there are two types of grayling—those that grow older because they know how to avoid fishermen, and those that do not. We humans know best those that do not. These are the ones we eat.

That Russian, a kind and gentle man born in the 1890s, was like most rural Russians. His knowledge about the natural world included far more than the behavior of fish. He was right in knowing that even among the same species, an individual's behavior develops by widely differing experiences. Some individuals may deal with violence in their past and become savvy. A fortunate few might pass through life devoid of serious challenge. Others succumb.

As innocent citizens of the United States have learned since September 2001, fear on a daily basis has crept into many lives. Nevertheless, some people and some cultures are more fearful, more vigilant, more wary than others. We as humans, however, differ little from other primates. Our priorities, like theirs, include finding mates, procuring food, and avoiding violence. Reliance on behavior as a means to thwart predators has been ingrained not only in our history but in our anthropoid ancestry and that of our other fellow mammals. But when beings of either the human or animal kind remain naive or non-savvy, they are subject to invasion—invasion by alien species, by hostile forces, or by other colonizers. If they fail to respond aptly, extinction fol-

lows. Mammoths, horses, and giant ground sloths, once common in North America, serve as testimony. So too do modern inhabitants of islands—auks, kangaroos, and tortoises. These losses all coincided with the arrival of a novel invader—a biped named Homo sapiens.

Were such species really so inept that alien invaders could not be thwarted? Do modern predators wipe out their prey?

That Russian—Harry Ordin, my grandfather—died 25 years ago. His saying about fish still rings true. If I had any hopes of being successful I'd need to avoid danger and learn from experience. But most of all, Harry told me to follow my dreams. This book is about a dream, a dream to study animals in the wild and to understand those that have and have not coped with changing environments, with predators, and with people. It is about their cultures and the lessons they can teach us about ours.

I'D BEEN SLEEPLESS for hours. Finally, the first tinge of light arrived. The eastern sky slowly shifted from black to an eerie pale. Below a mountainous horizon were treeless plains covered in snow and ice. The milling of hooves was not far off. There were elk—groups numbering in the thousands. Their fetid breath shot clouds of gray steam into clear, –20°F skies. This was Jackson Hole, a small valley bound by the majestic Teton Range in Wyoming. It had been almost three years since wolves were returned to Yellowstone National Park after a sixty-year absence. Three packs had come south. The year was 1998, the month, January.

The elk dallied and fed—a jumble of tawny fur that reflected the low morning light. Pawing craters to reach frozen grass, they barely looked up as two wolves approached. At twenty yards, an adult cow lifted her head, only to shake ice flecks from her well-muscled body. With nonchalance the lanky carnivores continued. Anxiously I awaited the elk's response. At 10 yards there was none. She continued to nibble. Then, she began digging a new crater in the hard-packed snow, ignoring the wolves. There was a sudden rush. No flight. The end came magnificently swiftly. Brutal. Blood and entrails stained the perfectly white snow red. A bald eagle flew in. Ravens were already on the ground. Three coyotes awaited the leftovers.

Astonished at what I had just seen, I shook my head. My frozen fingers scrawled the words "no marked reaction" in my data book. I added a question mark, circling it for emphasis. I was totally perplexed.

After a mere sixty-year absence, these elk seem to have forgotten a species that has made its living eating them for more than a hundred thousand years. I noted in my data book that this was my "first" mortality in a predator-naive population. More would come. I also imagined these elk as once having a culture deeply steeped in knowing how to avoid wolves. Perhaps the claims of imminent extinction by ranchers would be proven correct when more wolves reencountered more naive elk. I wondered about the fates of other species lacking experience with predators.

THREE YEARS PASSED before I again found myself sitting on what in 1881 Alfred Russell Wallace called "one of the least hospitable and bleakest spots on the globe," an island just ten degrees from the North Pole. This was Svalbard, Norway, in March. Ice crystals sparkled against sky as early morning light shown weakly on pockets of scattered caribou. In the –30°F air, steamy condensed breath jettisoned from frothy mouths.

This is a land where humans are but recent arrivals, and part of the Pleistocene lives on. The archipelago is so inaccessible that wolves and other land predators have never been occupants. Thousands of miles to the east, on Siberia's Wrangell Island, mammoths survived as Egyptians built their first tombs. On these Arctic islands, just as on the Galapagos, some species live without apparent fear, the caribou of Svalbard among them. They feed, pawing for lichens frozen beneath wind-packed snow, totally unafraid. Stunted in size, their bodies were engorged with fat. With no predators, there was little call for alarm, less incentive to run. These undaunted animals were fat and content. They still are.

Few caribou live like this. In Alaska they are neither runty nor do they maintain similar fat stores. They migrate up to 1,200 miles. They live with grizzly bears and wolves at their hooves. Whether culture differs among caribou societies or other species because of the presence of flesh eaters is simply unknown.

WHICH SITUATION IS the more natural—prey responding to predators or prey failing to do so? What makes for successful prey? Do we, as humans, have real choices about which landscapes should maintain dangerous animals? If so, do we betray the trust of species that now live in the absence of predation when carnivores are reintroduced?

During the last five hundred years, more than eighty percent of mammal extinctions across the world have occurred on islands, often among species with little knowledge of predators. Innocence can be fatal. But the most impressive incidence of how naiveté is rumored to have caused extinction stems from the loss of some fifty or more genera of large mammals at a point just after first contact with colonizing Paleolithic hunters.

If there are lessons to be learned and applied to conservation, we need to look both to past and current attempts to understand how carnivore reintroductions affect modern prey. If naiveté contributes significantly to the evil quartet of extinction—overexploitation, habitat destruction, invading species, and secondary extinctions—we must know why, and develop remedies to enhance the path to survival.

Conservation has a major goal: to hold onto what we have and to rebuild what we have lost. To understand how best to do this, we must understand animals, their natural history, and the threats to their existence. The science is not enough, nor shall it ever be. Real-world conservation victories are what count.

This book centers on three questions: (1) Can naive animals avoid death and population extinction when they encounter re-introduced carnivores? (2) To what extent is fear culturally transmitted? (3) How can an understanding of current behavior help unravel the ambiguity of past extinctions while contributing to future conservation?

The story I develop is entirely true. It begins with carnivores and prey familiar to most before moving in space and time to lands more distant and to species that stir the imagination. There are areas less traveled and past epochs we'll never visit. I also focus on a process—that of paths followed to pursue answers. This is because science does not and cannot derive purely from laboratory study nor computer simulation. Field work is required. My underlying goal is to impart the complexity of doing science under conditions that are less idyllic than we desire and more complex than we expect.

MOST JOURNEYS BEGIN with hope. Mine began as a pursuit to appreciate big animals in their natural world. I wanted to watch and enjoy their fascinating antics while learning how an understanding of their behavior and interactions with other species applies to their well-being. Conservation is not a sanitized field, and my account is as much a story about the realities of doing conservation as it is about the lives of animals and their behavior.

At times, my voyage was joyous and smooth, at other times, rocky. My ambivalence had little to do with physical challenges or work in remote lands. Instead, it centered on my fellow humans, on governments mired in hopeless bureaucracy, and on officials worried more about the next election than doing the right thing. Fortunately, there was also optimism, and this has led to successes. It is both, the despair but especially the triumphs that I elect to share.

part i

The Hunt for Eden

We live in a zoologically impoverished world, from which all the hugest, and fiercest, and strangest forms have recently disappeared; and it is no doubt, a much better world. . . . Yet it is surely a marvelous fact . . . this sudden dying out of so many large mammalia, not in one place but over half the land surface of the globe.

ALFRED RUSSELL WALLACE (1876)

chapter 1

The Wolf is at the Door—Who's Afraid?

> If they were not killers that show pluck and courage we should not admire them. . . . Great elk herds are not conducive to a balanced piece of nature without the wolf to add fire and alertness. . . . Herds of hoofed game, without the presence of a few carnivorous beasts whom they fear, lose much of their character and interest.
>
> EDMUND HELLER (1925)

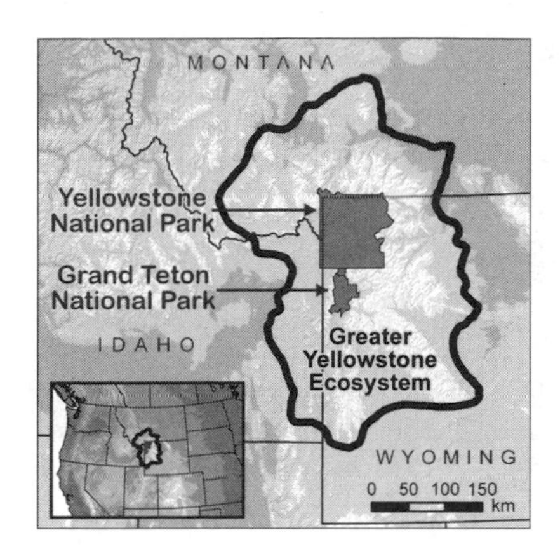

NOT ALL EDENS are created equal. To North Americans, Eden may be majestic forests of redwoods, or mountains of rock and ice. In Madagascar it may be where chameleons and baobabs flourish. South Americans might visualize a dusty Patagonia steppe. While a biblical Eden may be African, rich in greenness and warmth, with bountiful animals, Edenic places are world over. Whether paradise or frontier, not all are tropical. The North Pacific has its diving puffins and soaring fulmars, Antarctica its penguins, and Greenland its narwhals. The Eden of the past will not be the Eden of our future.

For some, it might not be a place at all. The Eden of the mind comes in many forms. It could be the wistful image of mammoths trumpeting their last calls across unpeopled landscapes. It could be memories of a past epoch when North America's savannas harbored the diverse wildlife of today's Serengeti. Or it might just be a system with its current biological wealth enhanced by the chilling howl of a distant wolf. Edmund Heller—naturalist, scientist, and hunter—mourned the destruction of wolves in 1925 in what many believed was, and still is, an Eden called Yellowstone.

The continuing global decline of large carnivores today has catalyzed great interest in reintroduction with the dual goal of restoring populations and reestablishing ecologically important relationships with other species.

The repatriation of predators like wolves and bears to their native ecosystems will always be an emotionally charged and culturally disruptive issue. Humans who incur few if any burdens associated with carnivores generally favor restoration; those living rurally and in closer proximity have greater antipathy. In the American West, carnivore repatriation has been especially divisive. Can prey species adjust to their presence? Indeed, can people?

JUNE 1973. Starry-eyed, hair long and unkempt, and with backpack in tow, I left California and hitched eastward for four days. At the site I had dreamed of for nearly three years, which I was finally traveling toward, are granite monoliths, volcanic extrusions, and calderas rising above prairies, quiet in their splendor. The air is stunningly fresh, not the smog-ridden skies of Los Angeles I knew. There are the sounds of cranes and curlews, marsh wrens and geese. Rivers rush, flowing clear and cold. Yellowstone.

Grizzly bears will soon be listed as an endangered species. Wolves have already been extinct for the past forty years. The charismatic megafauna of bighorn sheep and pronghorn, elk and bison move fearlessly. Enthusiastic tourists crowd the animals to snap photo after photo.

Yellowstone, the world's first national park, was dedicated by Congress 101 years before my unabashed and unannounced arrival at the office of the park's senior scientist. Fresh out of college—truly a neophyte in an arena I knew nothing about—I was hyper-alive, excited. I was intent on doing something of value. I wanted to study moose.

I did not know why—only that these giant deer are mammals of an ice-age epoch, of Asiatic forest origin, and recent colonizers of North America, having arrived less than twenty thousand years before me. But my rationale for study was elegantly simple and wholly simple-minded: less was known about moose than the park's other large mammals. Should I study their food or their feces, behavior or movements? I did not even know whether having a question—even a rudimentary one—was truly necessary. All I knew was that my heart was set on these Yellowstone moose.

My dreams were quickly dashed.

"Go back to California. People here get killed by grizzlies. They've been extinct for half a century in your state. Go to the Sierras, go home!" barked Glen Cole, the park's biologist.

In hindsight, it is clear that I should have followed Cole's advice. Instead, I ventured into the backcountry. As I hiked, I dreamed of watching wolves before their slaughter by government trappers—all within the confines of the putatively protected park. I saw the tracks of grizzly bears and became so terrified that I stayed up all night in my little tent, too afraid of being eaten to sleep. My logic was faulty, of course—I could have been killed whether sleeping or awake. I bathed in cold rivers and drank from clear creeks. I also managed to take a few meager notes on moose and learned that I could find and observe animals, and that I enjoyed it.

Then I tried to hitch back to L.A., but not before the local constable in Jackson Hole, Wyoming, escorted me to the limit of his jurisdiction: "There is no hitchhiking here, Boy—if you do so again, you'll be arrested." Little did I imagine that I'd ever work in the Yellowstone area. After my first experience, I was not sure I wanted to.

YELLOWSTONE IS HALLOWED ground for the global conservation community. From offices in San Francisco to Helena, Montana, and from Washington, DC, to Skukuza in South Africa, scientists, tourists, politicians, and the media note the events in Yellowstone. In a world where humans dominate, there are progressively fewer places within the earth's temperate zones where wildlife roam such vast landscapes. In Yellowstone, successes are legion and problems magnified.

Yellowstone National Park offers the world a concept of place and protection while underscoring the power of selfless achievement when the public and politicians work together. The park, dedicated in 1872, is managed by the National Park Service, whose early views of wildlife and protection differ from those embraced today. Wolves, extirpated in the 1930s, were reintroduced in 1995; eight years later, they numbered 175. Yellowstone is the best place to watch these exquisite hunters undertake the predation ritual as it has unfolded for eons. Yellowstone's doors also bear full witness to an array of modern problems—overcrowding, snowmobile controversies, and the slaughter of bison as they leave the park's protected borders.

The boundary issue highlights additional frailties, some associated with the Park's size. Despite its 2.2 million acres and a lack of fences, Yellowstone is too small to contain its long-distance megamigrants: elk, bison, and pronghorn. Beyond the park border, however, the ecosystem itself is huge—some 26 million acres—and largely publicly owned. This broad landscape makes it possible to envision a place of sufficient size where grizzly bears and wolves and cougars may yet interact with other wild species rather than coming increasingly into conflict with humans. This large, discrete area is now known more formally as the Greater Yellowstone Ecosystem. Locals call it the GYE.

The success of ecosystem approaches like the GYE has spawned protective actions elsewhere, such as Africa's Serengeti or Kruger national parks. But boundaries are just that, units fixed by political expediency rather than ecological appropriateness. The borders we deem sufficient for animals are mere constructs enforced by fences or bullets when animals leave. Where land is limited, so too is human tolerance.

By contrast, regions vast and remote, such as Alaska's Arctic Refuge, Tibet's Chang Tang, or Northeast Greenland National Park, have lines drawn on maps, but the formal boundaries really do not seem to exist. It is partially within these immense voids, areas indomitable by humans, that accounts of wildlife that tantalize the mind come.

I imagine species that do not flee. The prey may well recognize and be fearful of native predators, but in my Edenic dream, species are unafraid. Curiosity is the rule, not the exception. A fear of humans has not always been the case.

In 1742, Georg Steller, while serving as chief naturalist for Vitus Bering's explorations of the northern Pacific, described blue foxes—a color phase of the arctic fox—as unafraid of humans. He also contrasted sea otters from the heavily exploited Russian coastline with Alaska's Aleutians where they were not hunted:

> Lots of sea otters . . . I often found everywhere on the shore, which also shows that the (human) inhabitants, with enough other food, must not be much concerned about them, because otherwise they would not come ashore as infrequently as they do now in Kamchatka ever since the time when so many people developed such a liking for their pelts. . . . They could be seen (previously in Alaska) in large herds . . . playing or sleeping along the beach . . . as the hunting continued, they became more wary.

Sixty years passed before Matthew Flinders landed on Kangaroo Island, just off the southern Australian coast. Astounded by the compliance of western gray kangaroos, he reported the population to be lacking in fear; and because of "the extraordinary tameness they were knocked on their heads with sticks." Although these roos had lived without any predators for at least 2,200 years (an idea developed more fully in subsequent chapters), it was the interactions with dogs a few weeks later in 1802 that brought more public attention to the animals' naiveté. Francois Peron, a French zoologist, commented:

> With only one . . . dog, we caught in a few days, such a large number of big kangaroos that it seemed probable to us that a few such dogs, abandoned on the island would suffice to wipe out the race of these innocent animals.

Charles Darwin noted similarly while in the Galapagos in 1835:

> A gun here is almost superfluous; for with the muzzle I pushed a hawk off the breach of a tree. One day, whilst lying down, a mocking-thrush alighted on the edge of a pitcher . . . which I held in my hand, and began very quietly to sip the water.

Comparable observations are hardly confined to islands. In 1885, Teddy Roosevelt reported on change in the disposition of grizzly bears from the northern prairies:

> The introduction of heavy breech-loading repeaters has greatly lessened the danger, even in the very few and far off places where the grizzlies are as ferocious as formerly. For nowadays these great bears are undoubtedly much better aware of the death-dealing power of men, and, as a consequence, much less fierce than was the case with their forefathers, who so unhesitatingly attacked the early travelers and explorers.

These three species—sea otter, kangaroo, and bear—all seemingly shared in failed responses in their first encounters with armed humans. Lack of experience meant death.

Animals with hooves have proved no different. Two hundred years ago Lewis and Clark described wildlife as shy or reclusive when near human settlements:

> The whole face of the country was covered with herds of buffalo, elk and antelopes; deer are also abundant, but keep themselves more concealed in the woodland. The buffalo and elk and antelope are so gentle that we pass near them while feeding, without appearing to excite any alarm among them, and when we attract their attention, they frequently approach us more nearly to discover what we are, and in some instances pursue us a considerable distance.

Big game hunter Frederick Selous had a similar experience while bagging big game in southern Africa at the turn of twentieth century. "Where they have not been much persecuted, sable antelope are amongst the least shy of wild animals; and the bold and noble . . . gazing with curious though fearless eyes at the first mounted man to invade their haunts."

Even on Asia's central steppes, goitered gazelles were unafraid, a behavior noted by William Morden during his 1926 explorations. Sven Hedin, the famous Swedish explorer, went a step further. He used the reactions of wild yaks toward humans to judge how far outlaws might be from his encampments. Where yaks were not timid, Hedin reasoned they were plundered less by humans. Only thirty years ago, Mark and Delia Owens described how antelopes of the Kalahari Desert that lived in the absence of !Kung San hunters "acted as if they never saw humans. They played and jumped and frolicked and approached us."

Beyond these reactions to people, prey respond variably toward natural predators. Mardy Murie "couldn't remember ever having seen a moose gallop" in Wyoming, but in Alaska, where bears and wolves roam widely, moose certainly flee from predators. In a similar vein, desert antelopes of Namibia remained just beyond the bounds of arrows. Clearly, prey were learning about predators of both the human and nonhuman variety.

Is avoiding extinction really just a matter of learning how to fear different predators? If it is, then does the converse follow, that extinct species behaved inappropriately? Why did they not learn to fear as quickly? Questions like these are at the heart of arguments put forth by anthropologists and other proponents of the human-induced blitzkrieg of naive prey.

The idea of human overkill goes something like this. Many years ago, the meek fell to strange creatures. Timid prey made for easy meals, but a fortunate few avoided the feast. Over time they endured, not because of their cunning but because they simply out-reproduced their adversaries. Others adopted counter-strategies—hiding, fleeing, or moving. Some went to islands, some to deserts, and some shifted to the darkness of night. A few species deployed tactics more unusual—they grew larger or developed weapons.

The community of survivors had familiar names. They were rhinos and elephants that lived in the wilds of Africa and Asia, and vicunas and guanacos in South America. North America had its camels, horses, and mammoths, even antelopes, cheetahs, and lions.

A primate then stood upright, first in Africa. Over time, the hunting prowess of this biped improved. It entered Asia and Europe. Finally, it crossed the Bering Land Bridge to North America, where it encountered a fauna draped in innocence. Lacking deftness in behavior and failing in their cultural adaptations, the naive prey suffered a slaughter so colossal that nearly 75 percent of the species of large mammals perished. The extinct prey had been overcome by aliens.

Some think this story is true, others a fairytale. Unknown is whether the behavioral intrigues are plausible, let alone consistent with fact.

That some species survived has been at the forefront of arguments to refute the notion that naiveté rendered prey extraordinarily vulnerable to mass extinction at the end of the Pleistocene, some ten thousand years ago. Musk ox and pronghorn are modern survivors, as are caribou and mule deer. Only the pronghorn evolved in North America and without prior exposure to humans. So, as logic might dictate, pronghorn should not have feared humans and should not have survived. But they did.

An exception or two need not invalidate a premise, for answers are not always colored in black and white. Some species may retain the ability to rekindle ancient fears. Using domestic horses as one example, Charles Darwin suggested that, in the feral state, horses revert to behaviors that enhance predator avoidance. But Darwin also reported during his 1834 travels that the Falkland Island wolf was so unafraid that the gauchos killed them "by holding out a piece of meat in one hand, and in the other a knife ready to stick them."

While interesting, anecdotes about lack of fear do little to tell us how far back in time we need travel to understand species responses. Just because tigers have been absent from Pakistan and Kazakhstan for a hundred years, have Asian deer and antelope forgotten them? If tigers were miraculously restored, would their prey remember how to avoid them? No one of course knows, but Aldo Leopold considered the idea while searching for jaguars in northern Mexico in 1922: "We saw neither hide nor hair of him [but] no living beast forgot his presence, no deer rounded a bush, or stopped to nibble without a premonitory sniff for el tigre."

The spirits of North America are many. Jaguars are gone from Texas, Arizona, and California. By 1900, wolves had been extirpated from Maine to Iowa. Grizzly bears were purged from most of the American West by 1922 and from California by 1925. Some were big and bagged for meat, others were lost due to habitat destruction, and some were just in the way. In 1760, bison and elk were slaughtered in Pennsylvania and soon extirpated from the Atlantic coastal states. There were also oceanic losses. In the North Atlantic, gray whales disappeared centuries ago. Stellar's sea cows, first described from the North Pacific in 1741, were extinct within 30 years

Such losses have as much to do with missing species as they do the effects of losing ecological processes. Have deer forgotten how to respond to jaguars? Without bison, do we find less landscape disturbance and fewer flowers in what were once active wallow sites? Without eastern cougars and wolves, white-tailed deer devour the underlying vegetation. Acorns and birds are fewer. The point is not just that species are missing. It is that some species are likely to have played more important roles in affecting ecosystems than others. Wolves may be one such species.

How far back in time do we go to pursue these ghost effects? Is thirty years reasonable, or three hundred, or thirty thousand? No one seems to know—the standard is arbitrary. But the question about the past has direct relevance to conservation.

If systems missing key elements cannot be part of our framework for evaluating what is "natural" or "normal," then attempts at restoration will become increasingly problematic. What, after all, should be a reference point upon which we gauge ecological change and the "naturalness" of communities? Although mammoths cannot be restored, the knowledge that species like black-footed ferrets, California condors, and peregrine falcons were only recently lost from some systems and can be restored will guide the pathway toward recovery.

CENTRAL ALASKA, 1994. More than twenty years had passed since my ill-prepared meeting with the senior scientist of Yellowstone. Fortunately, my experiences had grown. I had just finished three years in the African bush studying black rhinos with Carol Cunningham. Our 4-year-old daughter, Sonja, was in need of a less feral existence, and we were in need of a new project. It was time to tend to new questions, to new systems.

North America would be calm, less diverse, less wild. Many of North America's ecosystems no longer held big predators. Between Canada and central Mexico, poison and persecution had reduced grizzly bears and wolves to less than 2 percent of their former ranges. Millions of bipeds had replaced predators. Guns and all-terrain vehicles substituted for arrows, for horses, and for foot power.

If a single question had been on my mind, it had more to do with the shape of future ecosystems than present ones. Given that more and more people would have less tolerance for predators, particularly in westernized countries, I wondered whether it would be useful to understand systems where large carnivores had already disappeared, for these undeniably are destined for our future. Predator-free systems were not the issue. The challenge would come in identifying comparable systems where a suite of large carnivores still persisted.

MY ENTIRE PSYCHE focused on the glacier as our single-engine bush plane hurtled toward it. I was searching for areas still dominated by wolves and grizzly bears. Alaska's Talkeetna Mountains might be the perfect spot.

Although forest and endless miles of taiga obscured any hope of seeing wildlife at lower elevations, the mountains had miles of open tundra. Thousands of caribou migrated to and from these mountains, but the most precious resource today were forty radio-collared moose. These were not typical forest dwellers, like most Alaskan moose. They were high-elevation Talkeetna moose that favored tiny islands of white spruce dotting open tundra. These moose crossed wide spaces.

This aerial reconnaissance was my first attempt to view the area before seeing it from the ground. Thoughts of study logistics faded quickly as the plane banked hard and dipped again. Gravity disappeared. As the glacier neared, I tightly grasped the metal railing for support and touched the harness strapped across my chest just for reassurance. Then we began a new set of torques. And, another. Beads of sweat formed on my forehead.

My stomach did not improve, although the glacier was now behind us. We had just sampled our sixth radio-collared female. Thirty-four to go. This was going to be a long flight, perhaps too long.

I was with Ward Testa, who had generously offered a seat on his flight to census moose calves. Ward, a state biologist, was trying to determine how predation affected moose populations. My interest was in how moose eluded the carnivores.

The relationship between prey and predator is not as straightforward as some might anticipate. Just because a predator lives in an area does not mean it is responsible for death. In Montana, for instance, up to 90 percent of the losses of domestic animals may not be due to predators, but instead to birth complications, abandonment, and inclement weather. And when native prey are available, predators may be more focused on them than on livestock. If Ward could muster the data on predation here in Alaska, I could channel my efforts more toward prey demography and behavior.

At this moment, however, I focused on other issues—actually, I could not focus at all. Again the plane plunged downward as we tried to flush the moose from the trees so we could see whether there was a calf. The apprehensive mother refused to leave the safety of her tiny spruce home. For me, flying sideways at ninety miles per hour, it was impossible to see if a newborn calf was cached. The plane swooped again, then again. I knew what was happening to my body, and the single thing I wanted most right now was to avoid puking. I reached for the air vent.

It was too late. My nauseating gags were amplified into the head sets worn by Ward, Carol, and the pilot Jerry Lee as the telltale bitter odor filled the cabin. Jerry made no attempt to calm my stomach or his flying; still needed were data on 34 more collared females.

During the next forty minutes I heaved only eight more times. Exhausted, my dry heaves produced an unanticipated effect: I passed out. The next thing I knew, we were on the ground, Jerry and Ward having called it quits after completing only half their mission.

Humiliated by my performance but undaunted in my desire to develop a project, I peppered Ward with questions about my objectives, logistics, and regions for potential observations. He suggested I contact a professor at the University of Alaska in Fairbanks, Terry Bowyer. I had known Terry when we were both graduate students, working in the deserts of California.

Although Terry was afraid of rattlesnakes, he studied bears, caribou, and moose. Together, Ward and Terry became my sounding board.

With my stomach still churning, we endured the Eureka Roadhouse at the southern end of the Talkeetnas. Wolverine, bear, and wolf hides hung from the wall. It was obvious that Carol and I were no longer in the lower US, where all but wolverines are endangered. Culture here was not exactly like Boulder or Boston.

In Alaska, the land is big and unforgiving. People do things differently. To our south was the Chugach Range, which runs unbroken to Anchorage. To the east was the Wrangell–St. Elias Range with Mount Logan, North America's second highest peak, rising to nineteen thousand-plus feet. Massive roadless areas, glaciers, deep snow, and brutal cold represented a first tier of problems in need of resolution. Others loomed almost as large—taiga and river crossings. And, if I wanted to believe Alaskan lore, then I could include grizzly bears and bush planes as other issues to conquer. Bear tales in Alaska were legion, and not a single person suggested that either my crew or I be weaponless. Terry suggested I sleep with a gun in one hand and pepper spray in the other, even in a locked cabin.

A final concern was entomological. Mosquitoes are widely known as Alaska's state bird. A Canadian researcher once received nine thousand bites in one minute when he stripped naked simply to conduct his experiment on blood loss. The snowmobiler seated next to us at the lodge listened intently to our conversations and then offered his view of our study plans, "What, are you fucking crazy—the bears won't kill you, the mosquitoes will."

Distances would be great, safety solutions expensive, and I had no idea if I would even be able to gather the right sorts of data.

To understand systems with and without predators, I would have to select sites other than just the Talkeetnas, including areas where ghosts of past predators might still be felt and others where they had long vanished. Some combination of regions with different prey species and carnivores would clearly be required. I wavered about my next step. Should I abandon moose for now and opt instead for caribou since they occur in more open areas where behavioral observations were possible? I could try for elk since they don't require planes for sightings. Or, should I drop back down to the lower US where moose live without grizzly bears and wolves?

Nagging at me were other issues, some occurring behind closed doors in Washington, DC, and in Yellowstone. These had not gone unnoticed by some, especially by livestock producers in the American West. Wolves were to be reintroduced to the park. If that occurred, I'd have the perfect opportunity to observe the results of a real experiment. The behaviors of prey species before and after wolves could be contrasted.

Soon after returning from Alaska, I jumped in my VW bus and drove to the Rocky Mountains. As I had done some twenty years earlier, I hoped to meet with the senior park scientist. Experience was now on my side. I had published technical papers in peer-reviewed journals about animal behavior and wildlife. That, however, was of little immediate concern.

At the moment, I hoped to talk about science, research design, and benefits both to the park and to society at large about my proposed study. I did not head to Yellowstone National Park. It was crawling with biologists anxious to get their hands on wolves. Instead, I went immediately south, to its sister unit, Grand Teton National Park.

The Tetons were the ideal place for those interested in naive prey. Not only had grizzly bears been wiped out, functionally rendering the park's largest mammals—bison, moose, and elk—predator-free, but grizzly bears were just now beginning to recolonize the Tetons from Yellowstone. If wolves were reintroduced into Yellowstone, they would surely reach the Tetons. I wanted to be there when that happened.

MY ENTHUSIAM FOR science and wolves was not universally shared. Throughout the Rocky Mountains, ranchers and the federal government had been hard at work for more than 75 years to rid the livestock industry of their perceived nemesis. Between 1887 and 1907, more than 20,000 had been killed in Wyoming alone. Despite their initial extirpation, wolves made it back to northwestern Montana by 1980. The nearest viable packs had once been in Canada's Jasper and Banff national parks, and it took more than 35 years and 150 miles of travel before their eventual recolonization of Montana's northern tier. The people who earned their living from the land would gladly wait another 35 for wolves to arrive in Yellowstone on their own, rather than through the hands of the federal government.

By 1994, some 50 to 75 wolves inhabited northwestern Montana. In the 14 years since their first arrival, the impacts of wolves on livestock were light—less than 40 documented kills of sheep and cattle or, on average, less than three per year. But the real issue was not one of livestock losses or safety or even the possibility of attacks on children. It was a battle about values in the American West—its land and ideologies, wilderness, and wildlife. It was about property rights and state independence, not that of a greater good or a benevolent goal. It was about local interests versus the will of America's resounding endorsement for restoration and conservation.

The brewing conflict over wolf reintroduction into Yellowstone reached far beyond the American West. As I looked for study sites in Alaska in 1994, the State Board of Game had just approved a plan to kill three hundred wolves a year. Aerial gunning was just fine. The goal was to grow more caribou and moose so people could then fill their freezers with meat. The plan, understandably, had tremendous support from local communities where subsistence hunting helped families through the winter and where grocery stores did not exist.

There was a different flavor to letters received by Alaska's governor and state fish and game department: "How I hate you all. You're crazy and sadistic. You're not hunters. You're barbarians. I will never visit Alaska."

GRAND TETON NATIONAL Park, Wyoming, 1994. The Office of Science and Resource Management was 2½ miles up the road from park headquarters in Moose. The building was ancient, and its roof sagged from too many winters of heavy snow. Now, in late May, snow was patchy, with some drifts still standing nearly two feet high. Just because we were in the lower US did not mean the logistics of Alaska would become obsolete. There were reasons why moose and bears and bison occurred here, and still do. The habitat was much the same as it was five hundred years ago, and winters were brutal. The lowest recorded ambient temperature in the Park was –60°F. Snow had reached the roofs of the vintage World War II dwellings that were adjacent to Steve Cain's, the chief scientist's, office.

Fully bearded, fit, and with red hair, Steve warmly greeted us. Sonja, still only four, sat in the corner as instructed. Surprisingly, she behaved. Steve put Carol and me at ease, letting us know his son was Sonja's age. As I described the study and objectives, Steve seemed supportive. Unlike Yellowstone's, Grand Teton's budget for wildlife research was nonexistent. But Steve felt that if we could find the funds, a proposal to carry out the research on park moose was likely to be approved. Elated, we left.

I worked on proposals, some aimed at gaining the necessary permissions from respective federal and state agencies. Because moose moved within and beyond the unfenced park, we would also need permits from the State of Wyoming. Other proposals were designed to gather data on whether predation on moose truly was relaxed. Although wolves and grizzly bears were not in the Tetons, cougars, black bears, and coyotes still existed. I needed to know if these smaller carnivores had an effect on moose survival. If they did, moose would not be predator-free, even though grizzlies and wolves were absent. The proposals were many and on topics that varied from predator-prey relationships to reproduction and behavior.

I planned to measure the responses of prey to the sounds and scents of predators. Some cues would be of the familiar variety and some of predators extirpated in the recent past. Other cues would be totally foreign. I now envisioned two broad experiments, one when wolves first arrived in the Tetons. Their appearance would enable a look backward and forward to understand how prey respond when they have not been exposed to a native predator for a period of time.

The other broad experiment would involve placement of sounds and scents to other prey species that remained either totally predator-naive or predator-savvy. There were to be appropriate scientific controls, sites that varied in predation regimes, different measures of behavior, manipulations that involved auditory and olfactory stimuli, and replication. Rarely did scientists have opportunities to perform controlled manipulations across large geographical scales. This was a dream project.

"Permission denied" said the voice at the other end of the phone. The Wyoming Game and Fish Department had rejected my request. "If we want *our* moose studied, we'll have someone from Wyoming do it." So said the forward-thinking director of the entire department. I now appreciated his moniker, Francis the Mule.

Although Grand Teton National Park quickly approved my study permit, and the State of Wyoming eventually followed suit, my funding was more precarious. Over the next two years, five proposals were denied, including some from ardent past supporters, the National Geographic Society and the National Science Foundation. This project was going to take work.

I needed to rethink my objectives. While it was obvious that the Pleistocene was over, there were still areas beyond Africa where the clock could be reversed. I could find places where time stood still—a hundred, five hundred, even ten thousand years. In the shadow of Siberia, there were wild regions where tigers still preyed on moose; little known spots where faunas from the Arctic and the Indian Subcontinent mingled. Not only did the world's largest cat coexist with grizzly bears and wolves, elk did too. Other sites held equal promise. Canada's remote Northwest Territories had wild bison that were the exclusive prey of wolves. Musk ox and caribou from parts of Greenland had never seen predators. In Yellowstone, wolves were at the door. Plans for their reintroduction were final.

chapter 2

The Shy Giant of the Forest

Who would have designed such a creature? It has none of the attributes of classic beauty; a face without merit; a drooping hang-jaw and protruding lower lips; a hard gummy pallet instead of front upper teeth. The moose has inspired none of the rhapsodies to the grandeur or sublimity of nature

DANIEL B. BOTKIN, *Discordant Harmonies* (1990)

FIGURE 1. Male elk in late spring. (Photo by J. McQuillen.)

FIGURE 2. Newborn elk calf.

FIGURE 3. Moose groups during early spring in Wyoming.

FIGURE 4. Sonja and Sage in spring.

FIGURE 5. Cabin at Square Lake, Talkeetnas Mountains. The craters in the foreground were dug by moose.

FIGURE 6. Female caribou in Denali National Park.

FIGURE 7. Raven in snowstorm. (Photo by F. Camenzind.)

FIGURE 8. Kalgin Island with the Aleutians on the horizon.

FIGURE 9. Aerial shot of a grizzly bear. (National Parks Service photo.)

FIGURE 10. Grizzly bear and twins in Alaska.

FIGURE 11. Antipathy to wolves is often vividly displayed.

FIGURE 12. Personalized license plate from Idaho.

FIGURE 13. Bison and elk at the National Elk Refuge in Wyoming.

FIGURE 14. Partially intact face of a female moose that had been alive three days earlier. Note the holes in her face due to predation by a grizzly bear.

FIGURE 15. Lu Carbyn (right) and the author on our improvised runway in northern Alberta.

THE TRACKS WERE large, three sets together. Each had four toes, distinct nails, and was anchored by a three-lobed pad. All were splayed in crusted snow a few inches deep. I first saw them on a blustery morning in November. They were next to an eddy coated in ice at the edge of the Snake River. The prints were unmistakable, twice the size of their coyote cousin.

It had taken 2½ years for wolves to arrive in Grand Teton from Yellowstone. Neither the public nor the park or state wildlife authorities knew they were in Jackson Hole. The primal hunters had yet to be heard or seen. Would their prey—moose, elk, bison, and mule deer—know the difference between coyotes and wolves, or were all wild canids just different-sized versions of each other?

Two days after I saw the tracks, the wolves crossed Snake River. A deer browsing on chokecherry fled indignantly. A moose lifted its head for a second look, nothing more. Elk, still migrating from the highlands, mingled among themselves on the sage steppes. The wolves stared, and a stalk turned to a chase as elk ran desperately in the dappled morning light. Enchanted tourists watched in the shadows of Le Trois Tetons.

Although this particular wolf pack would vanish for months, the newspapers said the experiment in predator-prey relationships was soon to begin. On that cold day in 1997, it already had.

***Dineega* means moose** in Athabascan, the tribal language of the Ahtna in central Alaska. To the Algonquin of Canada and New York, they are the "twig-eaters"—*monsoll*—or just *mooswa* to the Cree. As solitary dwellers of boreal forests, moose feed and move silently. They are the largest deer in the world, males exceeding 1,500 pounds, females 1,000. But size and mass carry a cost. Meat and other products are valued.

At sites on the Volga River in the Russian Urals, moose hides were fashioned for footwear and clothing more than 1,500 years ago. In Europe, moose raised as calves were ridden as adults, a practice that no longer occurs. Moose are still hunted for food across Russia, Scandinavia, Poland, and Germany, as well as throughout boreal North America. At the Pechora-Ilych experimental farm in Russia, 175 were raised for possible domestication.

In 1634, William Wood of New England described them as "not much unlike a red deare; this beast is as bigge as an Oxen; slow on foot, headed lik a Bucke, with a broad beame. . . . [T]heir flesh is as good as Beefe, their hides good for cloathing." Nearly three hundred years later, 14 moose would be transported to New Zealand, an introduction that ultimately failed.

Today, moose meat helps sustain northern hunters. Hooves, antlers, and fur are made into tents and blankets, tools, weapons, and art. Moosehead soup is served in some Athabascan communities. In Alaska, pellets are dropped from the sky during the "Nome Moose Nugget Drop." Tourists in Wyoming buy moose feces coated in plastic. Chic beers are named *Moosehead*, *Moose Drool*, and *Moose Juice Stout*. There is Bullwinkle, the cartoon character, the now-defunct Bull Moose Party of Teddy Roosevelt, and Order of the Moose, the Legion of Moose, and even a hockey team called The Moose. When a moose showed up in the deserts of northern Nevada it was shot. Incarceration was prevented by the outrageous claim, "I thought it was a buffalo." In Maine, 79 towns have "moose" in their name. In Alaska, fifty different sites are called "Moose Creek." For six years, my temporary home would be a town called Moose, situated at the foot of the Tetons.

Moose colonized Wyoming only within the last 150 years. Prior to the 1850s, trappers rarely saw them. But after fires scorched Jackson Hole around 1859, moose were more visible. In 1872, the Ferdinand Hayden Expedition shot one beneath the Tetons' highest peak, the 13,777-foot-high Grand. When elders of the Shoshone saw their first moose on the Wind River Reservation just to the east of Jackson Hole in the early 1900s, they were unsure of its identity. Populations initially grew slowly, but with modest protection coupled by the extirpation of wolves and grizzly bears in the 1920s and 30s, expansion was rapid. By the 1960s, winter densities of moose reached twenty animals for each square mile of river bottoms. Sportsmen around Jackson Hole have filled their freezers with this abundance, averaging about four hundred animals annually for three decades.

As in all mammal societies, females are more important than males since only females produce young. Males are the more expendable sex. Bull moose, with their huge palmate antlers, engage in fierce duels during the autumn rut that leave opponents bloodied, exhausted, and, when a tine pierces the eye, blind. When antlers lock and cannot be disengaged, both fighters die.

Although this sort of competition precludes many males from breeding, populations can always increase, because females are the limiting resource, not antlered suitors. Besides, male moose are harvested heavily outside of the Grand Teton park boundaries, the result being a skewed sex ratio and an exaggerated fright of hunters.

Females are easily distinguished from males. Even during winter when neither sex has antlers, one can identify gender just by seeing a butt. Cows have a white vulva patch about the size of a human fist, which males lack. But when the body is hidden behind alder or willow and only the ears arise, discerning the male from the female is not so simple.

Telling females apart from each other is also a challenge. Unlike zebras or lions with their individually identifiable stripes or facial whiskers, female moose look alike. Their ears vary little in size and shape, and their body coloration is often similar. Even their dewlaps—the little bolus of fur that hangs below the throat—is often indistinctive and changes from year to year. If I wanted to understand females and learn about their calves and relationships with predators, they would have to wear radio collars. I could then find and follow them in winter and summer, as well as learn about personalities.

That moose might each show unique qualities is an odd concept for many people. There is no reason why moose would not be similar to chimpanzees or humans in terms of individuality. Other animals are extremely variable in their behavior. Just as bison and baboons or woodpeckers and eagles vary individually, I expected moose to be no different. This variation, I hoped, would lend insights into survival tactics associated with the coming of wolves.

The park was quick to accommodate our request to radio collar 20 females. As the first nonagency scientists approved to handle big animals, sensitivity was requisite. Our efforts would be away from all tourists and backcountry skiers. A veterinarian was required. We would have to be in daily contact with rangers and the sheriff. Even the collaring material needed to be color coordinated to match moose fur. The park's mission includes visitor enjoyment, they did not want to degrade tourists' wildlife experience with visible collars.

With snow up to five feet deep, two veterinarians and I began our search for animals to immobilize. Alpenglow already coated the eastern sky where the snow intensified fiery hues. The air was calm, about 15°F below zero. Our packs had extra clothing and drugs, data books and darts, calipers and radio collars. There was also a receiver, tranquilizer gun, respirator, and tarp. To keep from freezing, lunches and water bottles were tucked close to the midsection of our backs. A thermos filled with hot water for tea would help.

The distant caw of a raven broke the silence of the forest. The holes we followed wound through trees of spruce and lodgepole and sunk deeply into the snowy surface. We followed a moose cow and her eight-month-old calf. Elk had all moved south or east. Our stealth was betrayed by the constant crunching of snowshoes. There were no moose to surprise.

Two hundred years earlier, Canadian explorer David Thompson commented on their wariness: "[the moose] is of a most watchful nature; its long large capacious ears enables it to catch and discriminate every sound; his sagacity for self preservation is almost incredible."

Our tactics changed. Although days still began as stars faded from the eastern sky, we no longer hiked to surprise a moose. Instead, we drove snowy roads until mule-like shapes emerged from willows or open areas. The direct approach worked wonders. Two weeks later, 11 females were wearing collars, each with a specific identifying radio frequency.

A pulse of forty pings per minute on my receiver indicated that an animal was alive. If the pulse changed to eighty pings, the animal was dead. Every collar had a motion sensor, so when an animal did not move for six hours, the sensor was tripped and the pulse rate doubled. The batteries that powered the collars would last for up to five years.

We needed nine more animals to reach the goal of twenty. Steve Kerr agreed to wait one day more before returning to his duties as a veterinary science professor in Colorado. Although we had darted animals throughout the park, I was anxious to deploy all remaining collars before Steve left. We would be forced to search for groups, even though the behavior of one animal almost always affected that of another. If one became vigilant or fled, others followed. Herds were not the circumstance I most favored, but Steve's imminent departure constrained our options.

Early on our final day, we discovered a group of 21 moose in snowy fields of bitterbrush and sage—nine females, six males, and six eight-month-old calves—all clumped tightly. We approached on snowshoes, sinking up to our crotches. Our luck held. The animals ignored us.

Steve fired the first dart, which settled with a smack in a well-muscled female rump. Four minutes passed. She became ataxic, switching from wobbliness to recumbent. We worked efficiently, whispering only when necessary. First, her eyes were covered, and a radio collar was slipped around her neck. Next, her condition was examined. The fur was dark and grizzled. Ribs were thick with fat, teeth in good shape, ticks few. Through palpation, we detected a fetus, knowing this mother-to-be was two-thirds of the way through a 230-day gestation. While holding her head gently, I removed the mask. Her eyes were gentle.

I designated her #010, and would offer a more respectful name once I learned something about her personality. While scientists are supposed to impart neutrality in labeling their subjects, it was far easier to know and remember individuals if they were named. This animal had no ungainliness—only simple beauty, sentience, and power.

As Steve finished working the immobile cow, I collected a sample of her feces. That way, I could subsequently compare levels of progestagens with her pregnancy status. In subsequent years, an animal's reproductive condition could then be assessed just by picking up and analyzing her feces rather than subjecting her to another traumatizing immobilization. I always preferred to use noninvasive methods of study when entering an animal's life.

I would still be required to locate each female, watch, and wait for dung, a sequence more challenging than it sounds. At times, the winds can be brutal, temperatures subzero, and the rivers that must be crossed ice-swollen. More difficult was when moose were in big groups, for the feces of one looks just like that of another.

With the handling of #010 finished, we neutralized the drug of choice, Carfentanil. Its narcotic properties separate the conscious mind from motor and sensory control. In other words, #010 was stoned and would remain so until her sensory neurons were blocked by an antagonist. Within five minutes, #010 was back on her feet, feeding and ignoring us. Only eight more moose to go.

We continued darting, Steve and I taking turns. The warm furry bodies that now lay anesthetized in snow were like people on a sunny beach in Cancun—unconcerned friends dreaming only twenty yards apart. My delusions about sun and beach were fleeting. Our fingers were raw, our toes frozen. By 6:00 PM, the temperature dropped back into single digits. Soon the sun would sink far west of the Tetons, and alpenglow would guide us back to the vehicles.

The next day tourists arrived, as they had the day before. The group of 21 moose had remained in the open snow-filled fields. "This is like a Gary Larson cartoon" a tourist offered. "One day all the moose are moose. The next, all are banded."

My sample now numbered twenty.

BY DEFINITION, PREDATORS kill and eat prey. Local outfitters had claimed that grizzly bears, now expanding outward from Yellowstone National Park, were killing most moose calves. They fretted because wolves were just around the corner. The claim was surprising, since juvenile survival rates had never been examined.

If calves were not abundant, factors other than bear predation might be responsible. Pregnancy rates might be low, and with fewer births there would obviously be fewer young. Winter starvation, disease, and even collisions with vehicles could all have the same effect. Sorting out the possibilities would require knowing if females were pregnant, if they carried fetuses to term, and if calves survived to or beyond their first winter. I would also need to know cause of death, something that troubled me, because only mothers, not calves, were being radio collared.

My first step, however, was to discover where moose went after departing their winter ranges. In areas of Scandinavia and Alaska, moose migrate several hundred miles between summer and winter sites. Perhaps in Jackson Hole, moose moved far during spring, maybe nowhere at all. They might just be concealed by vegetation. Even behavior could be involved. Females might be more apt to flee with new calves, or just remain if calf-less.

The valley known as Jackson Hole is big and foreboding, as it surely was to early trappers. Now it is a magnet for skiers and extreme sports addicts, ranchers and outfitters, and tourists. The valley floor is high—6,000 to 6,500 feet above sea level—and ringed by magnificence. The Teton Range is to the west, the 11,000-foot Gros Ventre Range and the 10,000-foot Mount Leidy Highlands to the east. Yellowstone's volcanic tablelands roll from the north. Other mountain ranges extend south, halfway to Utah—the Wind River, Snake River, Salt River, and Wyoming ranges. If moose moved out of Jackson Hole I would need an airplane and a pilot who knew how to radio track.

Fred Reed was perfect. He had flown in Vietnam, loved wildlife, and had his own aerial operation. His single-engine Maule was stored in the tiny hamlet of Driggs, just on the west side of the Tetons. My only hesitation about flying was my stomach, still tender from the Alaskan surveys.

My preference was always to locate moose from the ground. But when animals went missing, I would call Fred.

Eager to be airborne, Fred would coax his plane gently from the runway, gain altitude, and circle widely. My stomach liked that. We'd listen for the ping, ping, ping of moose in our headsets. Gradually, my aerial confidence rebounded. Fred and I took longer trips. We drifted across ranges of snowy peaks, looked beyond mountainous chasms to meadows glossy in morning light, and watched elk bathed in golden sunlight. We would see large black dots in forests: some were moose, others just our imaginations. Fred asked about my life, and I his.

Field biologists develop great respect for pilots, and Fred was at the top of my list. Then one day, during a tight turn, a strut collapsed. The Maule spiraled out of control and plunged into the ground. Fred was gone. The local community was stunned. Fred was the glue that bonded us all. At his memorial everyone agreed that Fred would have told us to move on. It was just something that had to be done.

"JB," asked Joe Bohne, the senior biologist for the Wyoming Game and Fish Department's local office, "you don't really expect moose survival *not* to change once wolves get here?" Burley, bearded, and with a dirtied cowboy hat, Joe was highly supportive of this study. He was also cantankerous and took pleasure in needling me, emphasizing the word "not" in his question. Joe suspected that wolves might play a large role.

With elk outnumbering moose by more than ten to one, I was less certain. I wondered why wolves would tackle a species as dangerous as moose if they could take elk. Like Joe and others, I did not have answers. That was the good thing about my work. I had no stake in a particular outcome. I just wanted data that I could impart on others.

In my pursuit to understand how the loss of predators affects prey and their ecosystems, I had assumed that where wolves and grizzly bears were absent, prey should be naive. If predation on moose in Jackson Hole was truly relaxed, then coyotes, cougars, and black bears should also have minimal impact on adult or calf survival. On the other hand, if young moose were being killed regularly by any predator at all, small or large, mothers might already be predator-savvy. If this were true, then the Teton study region would be totally inappropriate to address questions about prey naiveté and responses to alien or incoming predators. To do this study, I had to know if moose were predator-free.

Innocence among prey was not foremost on the minds of people living in the northern Rocky Mountains. They were more interested in whether wolves and grizzly bears were going to decimate the ungulates that lived in their backyards. Shreds of evidence from other ecosystems already suggested that novel but native predators might have such impacts. The information stemmed from carnivores whose ranges were expanding.

Cougars in the Mojave Desert were selectively preying on rare desert bighorn sheep at sites where cougars had been historically absent until about fifty years ago. The range of prey for the cats expanded only because mule deer had been introduced by humans in the 1940s to regions where bighorns lived. The cougars merely followed their primary prey, the deer, until they figured out how to kill native sheep.

A somewhat analogous situation has been unfolding in southeastern Quebec. Caribou have lived there for thousands of years. Coyotes arrived only recently—after the remarkable and swift transcontinental reduction of wolves. The small remnant caribou population in Quebec has continued to persist, but coyotes are now successful predators of calves, a situation that had not previously existed. Whether the naiveté of caribou to coyotes contributes to their decline is unclear.

The circumstance with Wyoming's moose differs in one critical respect. Wolves and grizzly bears had been predators of moose for thousands of years, unlike the recent events involving caribou with coyotes or bighorns with cougars. In many areas of Canada and Alaska, bears and wolves limit moose population growth. If Wyoming moose retained memories of wolves and grizzly bears, mothers might still be aware and cautious of predators, even if black bears, coyotes, and cougars had no impact on their survival or that of their calves. I needed to know if the moose of Jackson Hole were predator-free.

IT WAS MID-MAY and calves would soon be born. Thanks to their feces, I knew that 75 to 80 percent of the cows were still pregnant. But the more immediate challenge blended logistics and science, data and danger. Before birthing, females disappear into deep stands of fir and spruce or into brutally wet marshes and thickets overgrown with willow. They retreat to islands surrounded by engorged rivers, to impenetrable subalpine forests, or to islets of aspen surrounded by seas of sagebrush.

To find a newborn calf is a feat more easily planned than executed. Only the height of a full-grown cocker spaniel, just thirty pounds, and all legs and bone, calves are covered in a concealing reddish-brown coat. Most of their time is spent prone. If the base of a mother's legs is not clearly visible, babies will be unseen, a mistake that is costly. If a calf were missed and then died, a mortality event would be unrecorded, an omission that would bias the data toward greater survival. The bottom line was that I had to be on top of a moose to know if she had a new calf.

Joe Bohne asked how I planned to get in and count calves without getting killed. The hazard had been emphasized half a century earlier in a report from Montana.

> The cow moose with small calf not old enough to travel, will stand her ground. . . . Continuation of the approach [by a human] causes her mane to rise and one ear will drop down on her neck. When the other is dropped, she will charge and strike with both front feet.

The description of cows kicking with their front feet was memorable and accurate. At the time I read this account, a man had just been killed on the University of Alaska's Anchorage campus. The cow charged, smashing him to the ground with her chest. Then she used both her back and front legs, pummeling him savagely. Before he died, she lumbered above, refusing to leave, and allowing none of the horrified onlookers to help. To the north of Yellowstone, a bull once charged a photographer, driving the full fury of his antlers through the man's eye socket, killing him instantly. School kids, including my own daughter Sonja, have been kept from boarding their morning school bus by angry cows on more than one occasion. A different cow once poked her leg through the windshield of an Oldsmobile as a terrified owner sat inside.

From Wyoming to Maine and north to Canada and Alaska, moose attacks have seriously injured and killed people. *Dineega,* the shy deer of the boreal zone, had attitude, something unfamiliar to most who watch these apparently passive, awkward giants feed calmly on lily pads, browse on twigs, or glide effortlessly through deep snow. While most moose are peaceful most of the time, my quest for data would put me in touch with a different version of an animal. Joe was not quite sure how I intended to get accurate information.

If a calf was older than a day or two, it could move with great zeal, at a speed in excess of mine for up to a mile. In thick willows, I would probably never even see a calf and then the problem of estimation of the fate of a pregnancy returns. Even if I was lucky enough to sneak up on an unsuspecting mother, it still might not be possible to see her legs and know whether a calf was present. I could then assume there was no calf, deceiving myself into believing that females who do not flee have no calves. Or, should I assume that running mothers have calves? There were other possibilities; twins could be present.

To be certain of the accuracy of my data, I decided I could not afford to make any assumptions. I would just have to see a calf in order to consider it extant. If luck was on my side, I might observe calves from a safe distance. If not, I could be forced to move so dangerously close to an animal in thickets that I would be able to listen to her breathing. One thing I lacked was the luxury of being wrong.

There was an additional safety issue, one not considered by Joe or by the park. To get close to moose, I would have to hike in silence, a practice contrary to all recommendations in areas with bears. Several people had already been severely mauled by grizzly bears in both Teton and Yellowstone parks. To be safe in grizzly country, the idea is to make noise, dangle bells, and carry pepper spray. Outside the parks, outfitters and hunters regularly carried guns. This latter option was less desirable anyway, given that guns were prohibited in parks and my goal was to study wildlife, not mutilate it.

Nonetheless, my conundrum remained. Either I could make lots of noise, hoping to scare away bears, or hike in stealth, hoping to spot moose while avoiding bears. Worse yet, grizzly bears in late spring inhabited most of the same areas as moose—thickets, meadows, and bogs. When I conceived of this project in Africa, it sounded so simple, so domestic, so tame.

I HEARD THE familiar ping, ping, ping on my radio receiver affixed to the top of my VW van. My friend with the gentle eyes, moose #010, was in an aspen patch not far from the pavement. She'd always been unabashedly unafraid of humans. It was late May, and soon she would give birth.

I walked through sagebrush on a careful approach. A mule deer moved off silently; she had an udder, a sure sign of lactation and a fawn. Suddenly, a huge dark object rushed me. Her speed was explosive; ears were down, nape hair fully pilo-erect. Light momentarily reflected from an object on her neck. *Ahhh, the collar—number zero, one, zero.* This was not a bluff. It was the real thing.

I had no pepper spray, leaving it, in a moment of sheer brilliance, at the van. I hurriedly scanned for cover. None. The nearest trees were at the highway. For armament there was just nothing. The mad fury of hair and hormones was at fifty yards. A moment later, it was but twenty. My time was up.

I had one hope on my side—bluff and luck. I focused. I could not outrun this assassin as she neared for hoof-to-hand combat. I swung my lightweight radio antenna at her. I did it again and again. She circled me, stood for a few seconds, and began a new round of charges. Five swipes later, she moved off but only a couple of body lengths. I was in the trenches. The striking, hair-raised moose was not about to leave.

Still filled with adrenaline and eyes wide, I began jumping while holding my arms out, and grunting. It was not a move I had practiced. With each jump, I grunted louder and louder, like a monkey at the height of a tantrum. Primate or not, it did no good. She charged again. This time my antenna did something entirely different. It hit her squarely in the face.

I knew my game was now up. If this moose had an ounce of sense, she would have realized that she felt no pain. Then, from afar, I saw a tiny reddish shape emerge from the edge of aspens. The calf was still wobbly. *Man am I ever in deep trouble.* Mama looked toward her precious neonate. That's when bluff turned to luck.

I ran for the car, she for the grove. I didn't care where she was as long as it was not in my face. Shaken but inside the VW, I was relieved the relentless attack was mercifully over.

It wasn't. I looked out the window to see #010 headed toward the van. She followed my exact pathway, her nose fixed to the ground. Like a bloodhound on the trail of its quarry, this moose was on my scent. She crossed the road, and came to the door. She circled the van. A moment later she was gone, running back to the aspen grove and leading her unsteady calf away. The mother still needed a name. I now knew something about her personality. *Assassin* seemed perfect.

I had not known moose possessed the wherewithal to track odors over large distances. Ten years later I would learn about similar behaviors in an attack at Eagle Creek, Alaska. The five-year-old daughter of Grant Harris, a landscape ecologist for the US Forest Service, was out at the forest's edge near their home when she encountered a moose with her newborn. The angry cow ran over the little girl, returned to her calf, and then, using the odor trail of the retreating girl, back-tracked several hundred yards to Grant's home.

During the first few years of study in Grand Teton, I was almost always in the field alone. In addition to the fact that I rarely had funds to pay field technicians, I was afraid that a volunteer would end up maimed or worse. With time, my abilities to work silently improved, and I could get closer to moose mothers. I always reminded myself that it was not going to be the collared moose that "got" me, but the adjacent one. By thinking there was more than a single moose out there, which there were, my focus would be everywhere—not just in the direction of the radio signal.

One time, a charge came from a small tree island 75 yards distant. Although I again was in sagebrush, I managed to find one naked aspen tree for cover. As I readied my pepper spray, I reflexively looked at my watch, wishing for some unknown reason to time the encounter. When the irate female neared, I kept telling myself, wait, wait—wait until she gets closer before depressing the trigger. I could see eyes and nape as anger propelled her legs and chest forward. She had but one purpose in life—to end mine.

A red cloud exploded into her face as I unleashed the capsicum spray. Immediately she stepped back and began coughing, deep bursts resonating from her chest. It was not over. She turned and came at me again. I fired more pepper. Our standoff continued, her ears low, fur high, and me quivering behind the exposed tree.

Six minutes passed. She looked toward the grove from which she had come. I looked for my receiver, having dropped it twenty yards away. I tried to retrieve it, but she came again. Surprisingly, she now remained beyond the ten-yard range of the choking spray. *Moose could learn quickly.* My canister was almost empty. After another minute, she returned to her calf. I grabbed my gear, thinking I hated moose, despised this job, and wanted to quit the whole damn study.

A couple of days later, I talked with Steve Cain at park headquarters. Assassin had again struck. This time it was unwary hikers. She had attacked from a hundred yards, crossing a stream and charging at a full run. They were unhurt, but others had not been so lucky.

My field technician, Noah Weber, a strapping research assistant who had been an NCAA all-star lacrosse player, was staggered by a non-collared moose. She bowled over him but, like a Western gunfighter quick on the draw, Noah shot back. He deployed his pepper spray from the ground while looking up at her underbelly and udder. She backed off a little. Noah tried to run. But because of the thick vegetation surrounding him, Noah had gotten the dangerous capsicum all over himself. Unable to see, he ran blindly and impaled himself on a branch. He fell back to the ground, and still he could not open his eyes. Although the moose departed, Noah kept vomiting. Days later, his skin was still burning furiously and his eyes remained swollen.

IT IS NEVER easy to census calves, especially in deep willows or in marsh and bogs where thickets, water or mud always announced our intent. Perseverance brought progress. Sometimes we would work in teams, one of us pounding the bush and the other watching for a calf. Often, however, we had no choice but to plow through thickets, moving swiftly and decisively, a tactic that worked because mothers were not enthused about leaving their protected nests. We also recognized moose language. If a female looked behind her, it generally signaled the presence of a calf. But we still had to see it before claiming proof positive.

There were other frustrations. Thick vegetation blocked visibility, and despair reigned as sounds of moose shifted from faintness to silence as the animals fled unseen. Although we expected that an adult with a calf would flee, if we did not see the newborn, we would be forced to start all over the next day. Our censuses were conducted in rain or sun, snow or wind. There would be marshes and bogs, soakings in swift rivers, and insects. Damp or cold, it mattered little, for if moose were calving, data were required. I reached a point when I would curse the new leaves of aspens and willows, perhaps the only person in America who found the vibrancy of spring unwelcome. With leaves came obscurity. With obscurity came discomforting near-approaches to pregnant moose.

Field surveys did have their silver linings. On the floodplains we would see bears and newborn deer. There would be nesting sandhill cranes and trumpeter swans, the drumming of ruffed grouse, and screeches of red-tailed hawks. Meadows and prairies were carpeted in flowers—the yellow brilliance of balsamroot, the gentle whites of serviceberry, and the purple luminescence of sugarbowls. With long days of hiking, our bodies became healthy and strong. We also knew our moose. Some we labeled "runners," because they ran. Others did not. We knew personalities and began to document survival rates.

During the first few years of study, a period before wolves arrived from Yellowstone, ninety percent of the calves lived at least two months. About eighty percent survived to return to winter ranges. Such high rates of calf survival indicated that none of the potential predators—coyotes, grizzly and black bears, or cougars—had a serious effect. These Teton calves had about three times the survival rate of calves from Yukon or Alaskan sites where bear and wolf densities were far greater. If calves were not being affected, it certainly seemed that adults would be immune. If so, I would be on safe ground claiming these moose were relatively predator-free. Unknown was whether they retained memory of past predators.

GRIZZLY BEARS ARE the largest land carnivores in North America. Weighing up to 1,500 pounds, some exceed polar bears in size. They can outrun a horse. They fish for salmon, hunt musk ox, and scrounge for trash. They can smell the body of a rotting bowhead whale from six miles. For the most part, however, grizzly bear diets consist of grasses and sedges, insects and berries.

Grizzlies share much with people and with pigs. None have the specialized dentition of flesh-eating carnivores. They lack the well-developed shearing carnassials of wolves and tigers, and they do not possess the bone-crushing premolars of spotted hyenas. Although the jaws of grizzly bears are plenty powerful, like humans and pigs, bears are outstanding scavengers. All consume dead mammals, but their flattened molars are used more frequently to masticate vegetation.

As with most species, geography dictates diet. Coastal bears eat salmon. Central Alaskan grizzlies consume forbs but also caribou calves. In Yellowstone, white-bark pine seeds, army cutworm moths, and trout are the predominant diet. But there is one particular food item that separates grizzly bears in the GYE (the Greater Yellowstone Ecosystem) from bears elsewhere. Meat.

Yellowstone's grizzlies ingest more meat than others. This is no surprise, because winterkill—death by starvation—typifies elk and bison. Carcasses begin to appear in February and continue into April. Some winters produce more than others, but by early June, grizzly bears become hunters of young ungulates. In Alaska, moose calves are taken—one of the reasons for my choice of Alaska as a study site. In Jackson Hole and Yellowstone, there has been little evidence of predation on adult moose.

Through the first year of study, only two radio-collared adult female moose died, one by a vehicle and the other by starvation. I regularly heard the forty-beat-per-minute pings in my headsets of the females named Assassin, Belle, Sophie, Gypsy, Missing, and others. There would be calls from authorities of moose dead on the highway—seven once in a thirty-day period—or of an unfortunate dog stomped to death. But calls of problem moose were few, and reports of dead ones rare. I grew accustomed to the daily sounds of survival, the steadfast electronic pinging on my radio receiver.

That is, until one day late in May, when a mortality signal came from a remote part of the Tetons. The site was littered with dead trees and icy bogs. Most winter snow was gone, but shady forests concealed remnant drifts. As though it was an apparition bouncing off canyon walls, I chased the attenuated signal ringing in my ears for subtle changes in its pitch. When it chirped crisply, I was close to the dead moose.

Sinking above gaiters in soft snow while being mindful of an inadvertent moose, I stopped often. I hoped for the noisy chatter of ravens or magpies. Those were the sure signs of a carcass, just as are vultures clumped on an African savanna. There were tracks of elk, also of moose, some a few days old and melted. I wondered if these were of my collared female, an animal alive only three days earlier.

The Tetons rose like stalwart pillars glimmering in the afternoon sun. I emerged from the forest into an open field. An avalanche rumbled down a canyon, snapping trees as if they were matchsticks and sending a snowy froth high into the air above. Spruce dotted the banks of a nearby creek; the aspens were still leafless. A patch of coniferous forest was just ahead. The mortality signal emanated from behind.

Despite good visibility, I walked wide but noted something amiss. The snow was disturbed, and pine needles were strewn about. I unholstered my pepper spray and unlocked the safety. No longer interested in the carcass, I now looked for bear signs. I walked in circles—wide ones. Nothing. I crossed the creek, searching in the mud for tracks. Nothing.

Then I saw it: a deep print about eight inches long and half as wide. There were five toes—human-like, except that each had a detached singular depression. Claws. It was a grizzly, distinguished from black bears by its size and the straightness of its toe pads. My sounds became deliberate: "hey bear, Hey Bear, HEY BEAR!"

The pinging continued as I walked further into the closed canyon. Large rocks obscured visibility. More tracks appeared, then a bloodied vertebral column, a femur, and finally, a shredded collar. Fifty yards further was a skull. Also drag marks. No—these were not drag marks, they were signs of an active struggle. Branches the size of my arm were snapped. A few dwarfed lodgepole pines were crushed. A hair trail extended for more than ten yards, the site of apparent attack. Had the moose starved to death, there would be no evidence of defense, no signs of a struggle. Peaceful coexistence this was not.

I examined the skull. Two large holes pierced the maxilla. The distance of their separation matched the jaws of a male grizzly. The eye orbit was collapsed, and four inches of nasal bone were gone. The mandible was partially crushed. This adult was fresh meat, the first we knew of in the Tetons.

I felt sorry for the moose, a gentle mother with warm eyes. But grizzly bears also had a right to survive. In the lower US, they were an endangered species. Moose were not.

Whether this single death would clearly compromise my assumption that moose were predator-free was uncertain. A sample of one does not invalidate a generality, but I needed to know more about overall survival and develop criteria to evaluate whether prey were savvy or naive to predators. For most of the first few years of the study, adult moose survival paralleled the high survival rates of their calves. Exceptions existed.

Once, I tried to find the collar of a female that had gone missing. It was early June, and as before, winter snows had made river crossings dangerous. This mortality had been detected on one of our aerial flights, so I had a decent idea where the carcass lay—about ten miles beyond the park's northeastern boundary.

I hiked alone, shirtless and wearing shorts. My pack contained a medical kit to necropsy the carcass and other gear. Binoculars hung over my neck, as did the clunky radio-receiver and a morass of cables, wires, and straps. One connected to the antennae. Pepper spray was holstered at my waist, and I was doused with gobs of sunscreen. I was a man with a serious mission: collar retriever and moose mortician. My friends had their own description—geek.

My attire did not reflect that of a stud, something the cowgirl peering down from atop her gelding made perfectly clear. Dressed in denim, cowboy hat and silk scarf, and with boots made from alligator, this woman and her horse blocked my path. As I noted her gun, she told me it was for protection from bears. I kept going. She'd be the only person I would see.

Four hours of bushwhacking later, the loud pinging was finally in my ear. *I'm close.* There had been elk tracks, moose tracks, and bear tracks. As I climbed what I hoped was the last ridge, a bloated and surging river came into view. The radio signal was coming from just fifty yards beyond. I tried wading, but the water became too deep and the current too swift. Suddenly, something moved under a tree on the other side. It was dark. *Maybe the moose is alive. No—can't be, the signal is still on mortality. If the moose were alive, the pulse rate would be forty, not eighty. What is it?*

The hump moved again. Whatever it was, it had small round ears and a dished face. *A grizzly.* My heart began racing, this was too close—river between us or not. I left.

Two weeks passed. I returned when the river subsided. With me was Hank Harlow, a professor from the University of Wyoming. This jaunt with him was welcome company for me. Using sturdy branches as crutches, we clawed through the cold current and rapids to reach the female. Though her leg bones and jaw were scattered, a few deductions were still possible. Malnourishment was rejected. Her bone marrow was filled with fat. Trees were knocked sideways. Her skull experienced immense damage.

Hank, with his thick gray beard and deep glistening eyes, queried me.

"So, adult moose are not totally predator-free?"

"Yup. We've documented a few kills and these are up here closer to Yellowstone. To the south there haven't been any. All the collared adults are surviving."

"What are the causes of mortality?"

"Last winter, ten animals died in two and a half weeks. Not all were known as individuals. The snow was deep and many used plowed roadways for convenience, just to get from one place to another. Seven were hit by cars. Three calves had simply spent all their energy and died within days of each other. I'm not sure about disease. When we collar animals, we immobilize them and screen for brucellosis, tuberculosis, and other evidence of antibodies for disease. Nothing seems debilitating, but what we see when moose are alive might differ from what kills them."

We hiked back, discussing mortality. Despite the few kills by grizzly bears, more moose had been killed by vehicles, legal hunting, drowning, and entanglement in fences. The moose of Jackson Hole seemed relatively predator-free, a trait shared with moose from Idaho, Montana, Maine, and New York. Predation from nonhumans was rare.

BY THE TIME wolves crossed the Snake River late in 1997, the public was anxious to know when the first kills would show up in Grand Teton. So were my crew and I. By May, I had assembled several biologists to examine whether moose browsing affected the diversity of birds and vegetation along floodplains. The team included Peter Stacey, a distinguished ornithologist, and two graduate students. Soon after their arrival, they discovered the bloodied body of a moose calf. Paw prints and a trachea were nearby. If their suspicions held, this would be the first kill in park history.

I called Steve Cain and, together, with Carol Cunningham, we converged. Sonja, now eight, and Steve's son, Taylor, were told to remain at the car for safety. Bears had been sighted. Two ravens departed the carcass. Everyone was armed with pepper spray.

Steve and I searched first for bear, while Carol confirmed the wolf tracks. Then, my heart missed a beat. Two shapes approached the carcass—the kids. By the time we got there, it was clear bears had not yet visited, and the young sleuths announced their diagnosis. Whereas Peter and crew had reported a dead moose calf, Sonja declared otherwise.

"Dad, this isn't a moose and it's not a calf. It's a big fat elk." We knew then that Peter and his crew would be better off sticking to birds.

WELL BEFORE THE death of this elk, public tension over wolves was palpable. Tourists wanted to see more, as did many local residents, including some ranchers and hunters. The enthusiasm stopped there. Those whose economy depended on elk—outfitters, hunters, and the Wyoming Game and Fish Department—had concerns.

About 7,000 to 11,000 elk migrate each fall to the National Elk Refuge just south of the Teton Park. They come from adjacent mountains and valleys, some as far as Yellowstone National Park, sixty miles north. Several thousand travel east, where they concentrate for the winter in the

Gros Ventre Range. These elk all share one thing. To survive, they're fed pellets of alfalfa for several months each winter. Conflict occurs at these cafeteria lines, because wolves also enjoy a free meal. That is, elk. It mattered little to wolves whether the elk were young or old, because initially the elk treated wolves like coyotes. They ignored them. But on one cold wintry day, a crucial event heightened the fears of those who supported the feeding of elk.

The scene was carnage—bloodied, uneaten calves strewn across two hundred yards. Six, all killed the same night. A few bites were taken from a couple, but this meat was now destined for the bald and golden eagles, coyotes, and ravens that waited for us to leave. These calves would never grow large, nor would they ever find their way into a freezer. They would never shoulder the weight of trophy antlers or make more babies. These were dead elk, elk killed by wolves, and elk that rotted for scavengers.

Anti-wolf factions argued that wolf lovers had wrongly claimed wolves were not surplus killers. The dead calves contradicted that allegation. If this sort of thing were widespread, the elk population would plummet.

"Surplus killing," as it is formally known, has also been observed among other wildlife—wolves killing caribou calves and white-tailed deer fawns—as well as coyotes (and wolves) taking out flocks of domestic sheep. It's a behavior not restricted to North America. Spotted hyenas in Africa have killed Thompson gazelle fawns en masse. Surplus killing evokes great emotion, sometimes fascination, other times abject horror. While rare in occurrence, the few documented cases are held up to showcase what has been labeled "blood thirsty" behavior.

One potential trigger is the availability of easy meat—deep snow, vulnerable young, or prey maintained at high densities on feed grounds. The situation is no different from a fox in a chicken coop. The slaughter of fenced, domestic livestock is never pleasant. It affects human economies dependent on meat. Finding middle ground—the interests of elk hunters, wolf lovers, and people using public lands—is never easy.

"FEAR" MEANS DIFFERENT things, but its raw elements are similar—*alarm, disquiet, danger, panic, dismay, terror*. Because animals can't talk, we can only infer actions that may reflect fear. The measure of stress hormones is one such way. Changes in heart and pulse rates are another. But the most immediate and perhaps simplest way to judge fear is by behavior.

Few question that the behavior of animals, directly or indirectly, purposefully or inadvertently, signals intent. A growling dog bearing its canines sends a different message than one cowering with its tail tucked between its legs. A cat at full height with its back arched, fur erected, ears flat, and mouth open offers another indisputable sign. But as we humans become less familiar with wild animals, particularly species evolutionarily removed from us, such as rabbits or shrews, our ability to interpret behavior is compromised.

Fortunately, the prey species I study share behaviors when they respond to dangerous predators. The most overt is flight. More subtle is running to join a group and forming a tight unit just prior to fleeing. People do the same thing—we huddle together. We form tight groups if danger appears in a field or forest. Antelopes will jump up and down, a behavior known as stotting or pronking, while monitoring the movements of a predator. Even bison, despite their massive bodies, occasionally stot when alarmed.

As expected, individual animals are likely to be more attentive when danger is close. They break off whatever they are doing to listen, smell, or see to assess the degree of peril. If solitary, reliance is left to oneself. If a member of a herd, they may rely on the collective experience of others.

The different ways in which moose monitor their environments interested me. The first is vigilance, when a moose stands with head upright and ears forward. I would make a record of it by noting the degree of pulsing in its nostrils when one actively smelled its environment. Finally, I would measure whether a moose stayed or fled, a sign of perceived vulnerability to danger. I also planned to gauge whether moose were aggressive, noting retracted ears, the pilo-erection of nape fur, and the distance at which moose fled from me. With such information, I would contrast moose from different areas—where there was hunting by carnivores, hunting by humans, or no hunting at all. My expectation was that the type of predators with which moose coexisted would shape their behavior.

My primary goal was to determine whether fear varies geographically. If so, the question becomes why. The working assumption is that predator-naive individuals are less fearful than predator-savvy counterparts. If true, and either naive moose or elk fail to learn, then the claim that they will be decimated will be correct. On the other hand, individual elk or moose that acquire, retain, and use knowledge about predators to promote their survival will, by definition, learn.

Another possibility exists to account for prey that successfully avoid predators. Because populations are composed of individuals and not all individuals are identical in their behavior or other attributes, some types or forms might enjoy greater survival than others. Some personalities, for instance, are more likely to be curious of predators rather than be frightened to the point of fleeing; the bold may be more destined for removal from the gene pool. If so, the remaining members of a population will be the wary. Natural selection is the mechanism by which genes are propagated at different rates in subsequent generations.

To examine these ideas, I gathered information not only for Teton moose but also on those from Alaska, where life with bears and wolves has been continuous. Moose have other predators, notably Siberian tigers in the Russian Far East. However, my broader goal was to use moose as a stepping stone to ask similar questions about prey-predator dynamics and the transmission of fear in other species.

For the time being, however, moose remained my sole focus. In Wyoming, I was intent on determining how naive they were. In Alaska, moose live both on the mainland and on islands, sometimes moving in between. Whether their behavior differed based on geography was unknown. What was clear is that moose are good swimmers, navigating to islands in the Pacific more than a dozen miles apart.

Not all make it. While Alaska's southeast panhandle has fjords, glaciers, and brown bears, the ultimate predator hides below the water's surface. Although it comes in various sizes, there are common properties. All are dark with flashing bits of white-tipped fins. Jaws are filled with rows of honed teeth. They surface like behemoths before crashing back into the sea. A pack of wolves might be dangerous on land, but for a swimming moose, a pod of killer whales is deadly. One island-hopping female moose was proof. She never made it to shore from the straits near Glacier Bay to the north of Juneau.

chapter 3

A Tropical Primate in Alaska

Everywhere consists of amazingly high mountain ranges, most of whose peaks are covered with perpetual snow. . . . [I]t is my opinion that from here to the north as far as 70 more degrees latitude, there is nothing but land.

GEORG WILHELM STELLAR (1741)

ALONE AND CROUCHED, I caught a glimpse of the man's face. His skin was weathered, cheekbones high; he was handsome. His brown eyes burned with a deep intensity. His jacket was stained with blood, and tufts of snow clung to a fur hat. But it was the rifle of this Athabascan hunter that grabbed my attention.

In this part of Alaska, north of the glacially fed Knik River, moose avoid deep powder by dropping out of the mountains. The man shouldered the gun, steadying it with his left arm, much like a mother cradles a baby. Peering through the scope, he squinted ever so slightly. I swung my binoculars in the same direction. Silhouetted against the sky some 350 yards out was a dark object. Slowly the hunter removed his gloves and took deliberate aim.

The moose browsed unwarily. Spruce loomed above saplings of birch. Creeks were frozen. A bald eagle roosted in a lone cottonwood. Nothing in its environment was unusual. All was calm.

The moose moved two steps. Only its rump faced the hunter. Another second passed. The hunt ended abruptly.

The quarry had walked one pace further. As if an apparition, it disappeared into a lattice of trees. What was bad luck for one life was good fortune for another. This moose would live another day. Any further stalking was unsafe. Commuters to Anchorage roared past on the highway in Matanuska Valley. Sidling from bushes, the native hunter looked at me. I walked over.

"What are you doing?"

"Looking for moose."

"But, you don't have a gun?"

"I'm trying to find females. I want to play the sounds of ravens at them."

"Ravens?" he exhaled a deep sigh along with a look of exasperation, "What do you mean 'play the sounds'"?

I was straight but spared what, at the moment, seemed like ridiculous details. I expressed an interest in understanding relationships between moose and ravens. The topic changed as the hunter explained it was in his culture to put meat on the table. I asked what he meant about it being in "his" culture.

The man looked down in silence. Slowly, he removed his hat, shaking away tiny snow flecks. It was quiet, but not tense—an impasse—as if time stood still. I was comfortable. I looked toward the woods, pushing my hair back over my shoulders, and he began to describe his culture as one of the land and with the land. I struggled for a response that would ease the situation. I had no answer.

He tossed his pack into a battered pickup. Fuel barrels sat empty in the bed, one with an Exxon sticker, another BP. The expensive rifle was set gently on a gun rack. Cassette tapes were scattered on the seat. We nodded—an acknowledgement of unspoken differences, and parted, two people looking for moose. His goal was food, mine observation.

I pondered his comment on culture. It's obvious that we differ. In 1721, Pierre Charlevoix wrote that "The Indians look upon the 'moose' as an animal of good omen . . . .those who dream of them may expect . . . life." Little has changed today. Moose and humans are still intertwined. Some four hundred miles northeast of the Knik River, indigenous Chalkyitsik hunters harvested 35 moose in 1969 to sustain their Gwich'in village of a hundred people. Twenty years later, Catherine Attla, also of Athabascan descent, wrote, "When we get a moose, we hang up on willows or trees the parts we're not going to bring home. . . . [W]e do this to show respect, so the spirit of the animal will return to the land."

So, if the culture of the armed native Alaskan who I met was steeped in hunting, what about mine? My roots were European, a past that blended agriculture with wild game. I was influenced by parents who raised me with two brothers. Outside our doors were millions of others in the hybridized City of Angels. Ethnicities were diverse—Spanish and Mexican,

African and Asian, European. Most were now Americans. All were products of environments far from the boreal forest.

I knew nothing of this indigenous hunter, but guessed his past, like mine,was also linked to immigrants—his arrival from Asia predating mine from Europe by 10,000 to12,000 years. Still, relationships between Native Americans and wildlife have been very different from most Caucasians. In our ancestral histories, the distinctions blur. All were once hunters and gatherers. Before that, we were all potential prey. We are still a single species—a mixed and genetically interbreeding group of animals, *Homo sapiens.*

I WONDERED WHETHER my interest in ravens and moose struck the hunter as odd. Despite our differences, geographical variation and random events undoubtedly influenced our cultures and societies. Such variation was at the core of my interests. I expected carnivores to be the driving force that shaped how moose responded to ravens, an idea based on the relationship of scavengers with grizzly bears and wolves.

In areas where ravens do not make their living from trash or roadkill—such as northern Canada or parts of Alaska—densities are low. Food is limited, and temperatures may fall to –60°F. To survive, ravens have an odd symbiosis with predators. They are reliant on bears, wolves, even foxes to open thick-skinned ungulate carcasses. In their absence, the raven's access to juicy or frozen internal organs is blocked because they cannot easily pierce solid integument. Ravens might wait for days without feeding. But when a carcass is opened they are quick to the kill site.

The symbiotic part involves hunters, some human and others not. The Koyukuk of western Alaska tell of ravens garnering the attention of hunters, then circling above and cawing at the prey to enhance the chances for a kill. From Alaska's panhandle, Haida hunters have described ravens diving, tucking, and turning in flight to expose deer. Near Denali, in central Alaska, a raven even revealed the location of a lynx to the famed naturalist Charles Sheldon. That was in 1908, and the cat was promptly shot. Grizzly bears have been attracted to areas with a flurry of raven activity, something noted in the 1970s by Frank Craighead. A few years later, a Canadian researcher revealed how wolf howls attracted ravens.

These sorts of accounts whetted my curiosity. Were ravens, predators, and moose connected in a bizarre web of consciousness? Elsewhere, some birds act as sentinels and express alarm when danger exists. In the Kalahari Desert, yellow-billed hornbills called when snakes were discovered. Why wouldn't something similar exist in the northern boreal forests?

Ravens regularly feed on carcasses with eagles, foxes, and smaller scavengers. They might also wait nearby for bears and wolves to finish eating or just join the feast. If moose were savvy to sounds of dangerous species, they might also be responsive to ravens, particularly given the connection between ravens and carnivores at carrion. That was the prediction I hoped to test. Perhaps moose would prove less adroit than I suspected. They might be totally oblivious of ravens or, even if aware, blindly unresponsive to their sounds.

Taking a broader view of the possible role of carnivores, I'd have an opportunity to explore both the behavior and ecology of moose. Places where large carnivores were not exterminated, like most of Alaska, would serve as controls. Areas where wolves and grizzly bears were eliminated, which, in contrast was most of the western US, would be experimental sites. Ravens offered an initial—and key—first test using moose. Other species would follow later.

To get answers, I planned not only to measure the feeding and vigilance rates and other behaviors of moose but also to manipulate their acoustic environments. I'd amplify raven calls from a recorder, using a large speaker to broadcast the sounds. As a control, I planned to playback the calls of a non-scavenger—a red-tailed hawk—a species common in the same environment. To determine whether the playback system works, my dummy sounds would be running water, wind, or insects—all familiar cues that presumably cause no discomfort or alarm to moose.

To represent current predators or scavengers, I would use the howls of wolves and coyotes. Past predators might have been hyenas or tigers, both known prehistorically from Alaska, the Yukon, and Siberia. To assess whether moose had some ingrained memory of their deeper past, I'd toss in the sounds of whoops and roars of these two lost predators.

It is not always clear what, if anything, respondents react to. So I would contrast the sounds of these current and past flesh eaters with something less dangerous but equally unfamiliar—the roars of howler monkeys. The call of this neotropical primate has acoustic properties similar to the modulations of wolf howls, sounds entirely novel to moose.

My expectations were simple. If moose associated specific sounds with danger, they should respond differently to the cues of howler monkeys, since monkeys have never been dangerous to moose, nor have moose ever heard them. Moose might, however, view howlers as dangerous simply because their sounds are unfamiliar. It might be more prudent to flee something strange rather than remain to discover what it might eat.

To find answers, I needed to solve several logistical issues. Auditory equipment would have to function under extreme conditions—temperatures of –20° to –40°F. It would need to be robust, capable of transport, and powerable from remote field locations. It would have to fit on small planes, be wired to operate fifty yards or more from me, and survive winter and spring, ice and mud.

A SNOWY WASH blocked our view as skis glided across an ice-packed runway. The tiny Super Cub was soon airborne. A checkerboard of conifers and snow melded into pure whiteness as the single-propeller plane guided us above the treeline. My captain was Don Deering, a former World War II bomber. At 74, Don was spry and seasoned—the archetypical pilot of the Alaskan bush. I knew there were old pilots and bold pilots, but with Don, I felt better knowing there were no old, bold pilots.

We flew deep toward the heart of the Talkeetna Mountains. I would soon be dropped on a frozen lake where a trapper's shack would serve as base camp. From there, my experiments with moose, predators, and ravens would begin.

The icy backbone of what was once called Seward's Folly radiated splendor in every direction. Denali—North America's tallest peak at 20,320 feet—was to the west. Mount Deborah and Mount Hayes, also part of the immense Alaskan Range, were north. Chiseled glaciers of the Chugach broke the southern horizon. Seventeen of the continent's twenty tallest peaks are in Alaska. The volcanic summit of Mount Sanford in the Wrangell–St. Elias massif was just east, poking above clouds at 16,000 feet. Thousands of frozen lakes stretched in-between. The Talkeetnas were small, less than 9,000 feet, but their glaciers cover an area 25 times larger than that of the better-known Brooks Range.

Expanses of ridges and rivers, uplifts, and glacial outwash unfolded below the Super Cub. Further out were green-clothed forests of white and black spruce, frozen muskeg, aspen, and poplar. In March, the operative words were simple—ice, rock, and snow. Despite light that increased more than six minutes daily, warmth was barely in the air. Ridgelines were still drenched in winter coats, the rivers ice-covered. A muted sun illuminated the tracks of moose; wolf and fox spoor were less common. Bears remained in winter torpor, their heads yet to appear in this peri-glacial landscape. I felt alive, lucky. In sheer rawness and beauty, few places matched this.

Don banked hard before landing. Snowshoes bungeed to the outside struts blocked my view. Since the FAA rarely policed this frigid section of the world, Don had learned a few tricks during his forty years of bush flying. As a consequence, I'd dangerously squeezed more than my allotted seventy pounds of gear into the tiny tail section.

The items needed to sustain me for weeks included more than food. Among them were a heavy 12-volt battery would power my sound system, which consisted of a JBL speaker and an amplifier. In one section of the plane were my sleeping bag, 12-gauge shotgun, telemetry gear, ground-to-air radio, and solar panels. In another were canisters of fuel, a stove, precious matches, and extra boots. So were pepper spray, a down jacket and parka, and an array of gloves and thermals. Stop watches, electronic adaptors, spools of speaker wire, a bike pump, boot liners, and a backpacking cook set made up more gear. So did three cameras and a telephoto lens, lithium batteries, and film—all squished into my pack. Rounding out the list was a GPS unit, spotting scope, binoculars, and books. I tucked pictures of Sonja and Carol in as bookmarks, along with a small stuffed animal. It was Sonja's favorite little moose and now my lucky charm. "Lucky" would keep me cozy at night.

Due to the volume of gear, food rations were limited. Breakfast would be two packets of oatmeal. A single bagel with cream cheese and a combination of pemmican, pumpkin seeds, or energy bar would offer a daily snack. This diet guaranteed weight loss. A daily fix of caffeine would help. The additional pounds of organic coffee and cans of condensed milk would keep physiological anxiety at bay. Dinners were to be rehydrated beans, rice, or vegetables combined with a brick of cheese. Coupled with tortillas and habaneros, I'd be warm. Neither refrigeration nor mosquitoes were a concern—everything would be frozen.

Square Lake burst into view. Beyond was the Oshetna River. It was iced so solid that the water that ordinarily ran below was forced upward through fissures, a phenomenon dangerous to humans known as "overflow." A ridge rising above, called Big Bones, radiated light. The skis touched down on the frozen lake.

As I ferried gear to shore, Don was already back in the air heading south. Like a tiny moth, the Super Cub soon disappeared, its drone turning fainter until the snowy landscape was deathly silent. With no wind, the only sounds were mine—snowshoes splitting fragile crust. I breathed heavily as the last of the supplies were lugged toward my new home, the hut. It had walls of thin plywood and lining of newspaper; the windows were large vacuous openings covered in plastic. My sleeping bag, rated to −40°F, was a good investment.

Outside the single-room shack were 55-gallon drums, castaways sitting empty as if littered from a passing ship. Rusted cables and chains mixed with half-buried bones of moose and caribou. Beyond, the refuse of modern humans quickly disappeared. Muskeg and dwarf willow merged with ridgelines and soon with mountains. Two miles away a solitary animal fed above treeline. Its distinct white vulva appeared large in my scope. I smiled—data collection could begin tomorrow.

I prepared for a cold evening. Dinner required water, which meant one of two possibilities. I could cut a hole in the 3-foot-thick ice, drop a pail into the lake, and haul the water back to the cabin. I'd have to do this with snowshoes and carefully, otherwise much would spill along the way. But even if I got it there, I'd still have to collect wood and build a fire to kill giardia. The alternative was equally fascinating. Wait for sun to melt snow packed on the metallic roof. Two feet remained. Earlier in the day there was a promising sound: a drip. If that continued, I'd not even bother with boiling. I could live with a few fine metallic speckles in my drinking water.

Soon, I'd have another visitor. Actually two. My first would be Kevin White, a biologist from California. He'd arrive courtesy of Don in a few hours. Kevin knew mammals and birds. He was also a patient observer of animal behavior, fit and self-sufficient. This would also be a good opportunity for Kevin to decide whether he wanted to attend graduate school or be a field grunt all his life.

A few weeks later, a second visitor would arrive. It would not be human. Coming by helicopter would be a 4-wheel all-terrain vehicle. Kevin and a another coworker who would be arriving later would use it to navigate muskeg and rivers in search of moose. Ultimately, it would be their ticket out of the wilderness, and back to the highway sixty miles distant at summer's end.

By African or Alaskan standards, our hundred-square-mile Oshetna study area was puny. A mere snippet of mountains and hills, swamp and river, it was only ten miles across and ten miles wide. A bush pilot would zoom past in the blink of an eye, an interstate commuter in ten minutes. Neither would be the case for Kevin or me. Our transportation network had two elements—the trails we packed by snowshoes and our ability to power along them.

Thirteen radio-collared female moose were in our Oshetna study area, some one or two hundred more without collars. Our immediate task was to climb the granite knoll west of Square Lake. The protrusion, less than a mile away, would offer visibility in all directions and improve our chances to hear radio signals.

By morning, the outside temperature dipped to –2°F; inside the cabin it was a toasty 8°. I managed to keep my three cameras, two cans of condensed milk, and some water from freezing by sleeping with them.

At 5:30 AM, Kevin still burrowed deeply in his bag. I primed my stove and flicked a match. Soon the lightweight heater was purring, blue flames shooting below a pan still thick with ice. I unscrewed a canister of coffee, deeply inhaling the rich aroma. Kevin's sleeping bag swiveled, and a finger slowly appeared before morphing into a hand and then an arm.

The stove whirred, warming the shack to 13°. In gloves, hats, and down jackets, we enjoyed coffee and oatmeal. A visit to the outhouse would have been next, but there was none. Instead, we post-holed in snowshoes to a spot we designated a latrine.

The condition of snow was everything—the difference between a good day and bad. When soft or lightly crusted, we struggled, sinking deeply, even with snowshoes. A mile of progress might require four or five hours. Despite the freezing temperatures, the hike would leave us coated in sweat, even after stripping down. Rising early helped. The snow was still hard, making the use of snowshoes optional. Other times, we would just glide as if running down a track. Once our trails were hard-packed, neither Kevin nor I wanted to veer from them.

While many biologists use skis to navigate the back country during their studies of wolves with either elk or caribou, I almost always opted for snowshoes. Given the propensity of moose to hunker in thickets of alder or willow, travel was slow, whether with skis or snowshoes. Perhaps I had just been imprinted on snowshoes, but I felt more efficient with them strapped on.

With packs and playback gear, Kevin and I looked for moose and gradually began to know the area. At first, the landscape all seemed the same—white and gray. Only the homogenized greenness of spruce added color. The silence was welcome—no planes, jets, or vapor trails. Although the rivers and creeks were still iced over, slight gurgles of running water grew daily. Winter would soon transition to spring.

The increasing light brought other changes. Willow ptarmigan, their white winter feathers matching the snowy background, became vocal. Males, with reddened eye patches, became interested in females. The sounds of "kohwa" and "aroo" broke the stillness. These were offset by more calls—"kok," "ko-ko-ko," and "krrow." Some 75 years ago, biologist Edward Nelson spoke of ptarmigan: "They move in flocks, often numbering several hundred, during their migration, when they pass to and from their summer haunts. Among the Alaskan natives . . . , especially those in the northern two-thirds of the Territory, this bird is one of the most important sources of food supply, and through the entire winter it is snared and shot in great abundance."

Other birds appeared as we plowed through snow—gray jays, flocks of redpolls, and chickadees. On one 10-mile day, we saw four bald eagles and a dozen ravens.

The rarest prize sat atop a small spruce staring downward. With brilliant yellow eyes and small white spots on its crown, this earless raptor was totally unafraid. Northern hawk owls are spread across the boreal forests of Canada, Russia, and Alaska. We enjoyed hearing its "kee-kee-kee" resonate and watched for almost an hour under a warming sun.

As our familiarity with the Oshetna region grew, we also found more animals. From hills beyond Square Lake we heard pings from radio-collared moose. We saw the tracks of wolves; twice, those of wolverines. One day we watched a porcupine floundering in deep snow about a mile away. It reminded us of us—a being deeply buried, struggling to move. With our spotting scopes focused, a different picture emerged. Quills disappeared as small round ears poked above a broad face. The animal was not struggling in snow. It muscled through it. This was not North America's second largest rodent. It was a grizzly bear—a blend of power and determination pushing forward. At this time of the year, carrion would be its prize, since vegetation was still dormant. Squirrels, fish, and insects were not yet available. With bears now active, the victims would be more than moose.

Aircraft, much like Don's Super Cub, are used for hunting just as they are for transport for fishing. During early spring, large snow-covered areas are searched for the tracks of bears. When found, human hunters are alerted by radio and arrive by snowmobile. Even without aerial support, squadrons of four to six snowmobile riders course ridgelines for days or weeks looking for spoor. Terrified, the bears have little chance of escape. A few lucky ones flee over rocky ridges that snowmobiles can't reach. More often they are shot at a distance, never knowing what killed them.

Bear mounts, bear hoods, bear rugs, and bear gall bladders are all fancied for status or money. Bear meat becomes sausage or burger. For some, killing a bear is a sign of manhood, but for others reasons are less personal, more altruistic. It is a way to increase the number of moose and caribou.

PREY AND PREDATOR have had a tumultuous history throughout much of the world. From where we sat in Alaska's Nelchina Basin, it was no different. Some seven times the size of Yellowstone, the Nelchina was once roamed by a few Ahtna—a group of Athabascan descent. Before Alaskan statehood, starting around 1948, the US Fish and Wildlife Service began systematically killing wolves in the Nelchina—using poisoned bait and aerial gunning. About two hundred died in three years. By the 1960s, moose and caribou populations irrupted, with caribou numbering 80,000 and moose 25,000. Within ten years, the caribou population had dropped ninety percent. These were the wild days of wildlife management—do what you will, answer to few.

Since then, populations have fluctuated, and some control of bears and wolves has occurred. Female moose have not been harvested in recent years. When Ward Testa began work in 1994 in Nelchina Basin, which included my focal region in the Oshetna, grizzly bears were twice as abundant as wolves. Bears averaged about twenty individuals every four hundred square miles. Moose calves had poor survival rates. More than seventy percent died within their first six weeks of life, most by bears. Wolves took a few but generally they exerted strong effects only during winter and early spring, when calves and adults fell through the crusted snow. Wolves, being much lighter, did not. Unlike the moose of Jackson Hole, both of the big predators—wolves and grizzly bears—were playing key roles in the Alaskan wilderness.

WITH PACKS HEAVY and snow deep, Kevin and I moved clumsily toward the first radio-collared moose. At two hundred yards she detected us but elected not to run. Looking up only once, she resumed feeding. Her calf continued to watch us. They stood five yards apart.

We removed the playback gear and placed it firmly on a plastic sheet, hoping to keep the snow from jamming the cassette unit. I removed a box of tapes, each labeled with the appropriate sound—raven, howler, hyena, tiger, wolf, water. I also arranged the frozen cables and connected the battery to the amplifier with alligator clips. Kevin prepared data sheets and readied binoculars and stopwatches. He was to record and write. I was the equipment manager. The air was calm, the only condition under which we could perform the sound playbacks. Once finished, we would approach the area where moose had been and measure other variables—the distance between the speaker and the subjects, snow depth and plant cover. Each was necessary, because if the snow was too deep or the vegetation too thick, moose might vary their responses accordingly. By generating a sufficient sample, it would be possible to determine how each factor affected an individual's behavior.

Would the mother run or stay, become aggressive or keep feeding? For how long would she remain vigilant? Would she change after hearing the sounds of a raven or a red-tailed hawk, a familiar predator or an unfamiliar one? What about the calf, would she learn from her mother or adopt behaviors entirely on her own?

Before beginning with the sounds, we recorded the feeding rates of the duo to establish a baseline rate prior to offering sounds. Each fed uninterrupted for 180 seconds. The sounds of running water were then inserted in the tapedeck and played for 22 seconds. The volume was adjusted to the same level for all playbacks. The mother looked up for nine seconds, the calf for twenty. Next came the caws of ravens. Within five seconds, the mother was looking. The calf became vigilant within two seconds of her mom. The mother walked a few steps, lifted her head high, and focused her ears in the exact direction of the speaker. She was no fool. Kevin and I froze, collapsing in the snow. At 22 seconds, the raven abruptly stopped calling. We continued with data collection on the "post-call" response. The calf returned to feed after 31 seconds, the mother at 52.

The calf was not mimicking her mother. She acted on her own. Otherwise, both would have used the same behaviors for similar periods of time. Such differences between the mother and calf would figure more prominently in my later search for culture as I pursued an understanding of how behaviors were passed, if at all, from one generation to the next.

I removed the tape of the howler monkey from the playback unit. The mother had listened for the entire 22 seconds—not worried enough to flee, but not so comfortable as to continue feeding. The calf ignored it. We waited the obligatory period for both moose to return to their normal behaviors—feeding, what else? Given the mother's size, forty to fifty pounds of twigs, stems, leaves, forbs, and needles were needed daily. Unlike lions or dogs, moose feed many hours a day.

It was time for the wolf howls. Kevin looked over and smiled. I winked, and hit the play button. The duo did not respond—a failure that lasted all of two seconds. Vigilance was immediately followed by flight. The mother ran first, but her calf was right on her stubby tail. The mother's ears were back, as she lifted her legs high and skillfully through the snow. They vanished from view after a few seconds more.

I knew we would continue doing playbacks for more moose. Next time, I would vary the order of sounds, just to be certain that it was not the sixth or the third or even the first sound that they were responding to. That way, I would know if moose were responding to specific cues rather than just an order of progressively different sounds.

By the time we hiked back to the cabin, the snow had softened and we sank and sweated. We also celebrated our first data. My mind flashed to Sonja and the playbacks we did together in the Tetons. When I asked the then 5-year-old what an experiment was, she surprised me by saying it was tricking an animal by making it think that something is there when it is not, and then doing it differently to make it think something else is there.

She was right: the deliberate manipulation of one variable while holding others constant. Maybe she'd be a scientist. Parental naiveté no doubt.

With only about a mile to go, we neared the Oshetna. The river had changed in a week's time. Ice had collapsed. Torrents of water surged forward like an explosion ripping through a canyon. We looked for a different spot. With each step across the ice, I poked my ski pole. I unbuckled my pack. If I plunged into water, I could at least wriggle free. As the shoreline grew distant, my confidence faded. Ice was thinnest where the current was swifter, precisely where I stood.

A loud sound—cracking and ripping—broke my concentration. Ice was fracturing in every direction, the cracks racing forward as if layers were unzipping. I stopped, expecting instantly to plunge downward. Anxious seconds ticked by but nothing happened. I continued my crossing.

Ten seconds later, my left leg disappeared, then the right. Frigid water surged into my boots and pants as I stared at jade green ice. Clumps rushed at me, hitting my chest. I clung to a fragile shelf of ice. Like a spider, I flattened myself, extending my limbs in all directions. My snowshoe caught on a ledge. I moved a hand backward to untangle my leg, hoping the remaining ice would not give way. Other than being wet and shaken, I was okay. Extra boot liners and socks were a necessity. In the future, they would accompany me.

Over the next week, our success in finding moose continued. Some animals had collars, others did not. We played our sounds to subjects from 75 to 700 yards distant. By the time we finished our eightieth playback, we were fried. Our faces were crisp from the intense sun, fingers and hands cracked deeply from the cold.

When we'd reach the cabin at day's end, our feet were often sweaty and our boots soaked. By the time morning arrived, our wet boots and socks were frozen, since the unheated cabin never warmed much above 20°F. We had no choice but to climb back into the frozen apparel as well as our greasy, smelly, sweat-stained clothing.

Most days were excellent for data, others for unadulterated Alaska. The trail we crafted to reach the knob behind Square Lake was now home, a familiar freeway. From atop, we savored hot coffee from our thermoses and watched moose. Once, a grizzly bear walked past an unwary female only a hundred yards away, each oblivious to the other. Another time, nine moose fed below. We played sounds without bothering to stalk closer.

On cloudless days, the Alaska Range and the Wrangell–St. Elias massif jutted above the taiga. But because it was often windy, we did sit-ups and push-ups or climbed into our sleeping bags just to stay warm while doing observations. Sometimes, we hiked. Once I discovered an old fire pit overgrown with moss and lichens. Nearby were an arrowhead, a stone tool, and shards of mammoth ivory.

During late April, something odd occurred on Square Lake. Weeks earlier we watched four swiftly moving objects crossing ice: snowmobiles. Later, when we encountered the riders, we learned they were hunting bear. They would have also accepted a wolf or two. Now, five novel shapes appeared on the frozen lake.

The newcomers moved slowly. Grabbing binoculars, we expected wolves, but these beings were buff with light rumps and were strung out in a line. Caribou had arrived. Over the next few hours, we watched as three hundred moved north, crossing the lake. Spring was truly in the air. The migration had begun.

On a different day, when Ward was out flying, we talked on the surface-to-ground radio. He described a cabin more substantive than our hut. Just four miles away, it was cushy and had heat, even an outhouse. I had never met the owner, George Davidson, who was from Juneau, but he had given us permission to visit. We checked it out, even borrowing a few morsels of chocolate. "The Hilton," as we termed it, had rugs, windows, and thick walls. There was artwork and furniture, even a kitchen. The outhouse had a throne with views of Alaska in all directions. A group of 22 moose fed only one ridge away. Wolf tracks were imprinted in mud.

NEARLY 12 MONTHS had passed since my first visit to The Hilton, deep within the Oshetna wilderness. I had written to George asking about using his homestead, and he generously agreed. So, when Kevin and I called to thank him late on a Saturday from Anchorage, we discovered George, too, was in town. He and his wife, Georgina, were anxious to meet.

Kevin and I had just come in from a month in the field, and the brunch invitation was at the Marriott. Our clothes were filthy, and we hadn't showered. The night before our planned meeting, we camped without a tent on the shores of the Cook Inlet. We awoke soaked, caught in a surprise snowstorm. A dozen Dall sheep grazed on cliffs above. Wet and cold, we washed our armpits in a grimy bathroom. I anticipated climbing into clean jeans for our morning date with the Davidsons. Then came a shock.

The pants didn't fit. They were three inches too short, and wouldn't button. I checked my bag hoping somehow it was the wrong one. It wasn't, and there was less than an hour until we were to meet George and Georgina.

I had lost weight, not gained it, and was simply mystified at my predicament. The pants were Carol's. I had grabbed the wrong ones in my hasty packing.

My two options were equally unattractive, wear pure filth or hers. The brunch was drawing close, and the list of invited guests had grown. With us would be some of George's friends; the event was a Republican fundraiser.

I had no choice but to "cross dress." My simple-minded plan was to keep George and Georgina in a face-to-face, eye-to-eye vice grip so they couldn't look down. With my shirt-tails out, I'd conceal my, I mean Carol's, unbuttoned jeans.

Other than being bearded, I didn't know what George looked like. Until I discovered who he was, I'd have to be in each whiskered man's face, a disheartening task, given the proportion of unshaven Alaskans. When a man in his mid-fifties arrived at the brunch with a woman, I introduced myself. I had found the right couple but still felt—and acted—odd. To make sure that Georgina could not see my shortened pant legs, I stood close, putting myself immediately in her face. Then, I rotated back and forth between the two. Brashly, I guided them to chairs. To my chagrin, they ordered the buffet for all four of us. As we stood, my clothing again became my bane. I walked behind George and Georgina so closely that the people at the neighboring table asked if I was a shadow. We all sensed a growing weirdness.

By the time we sat down to eat I came clean, explaining my behavior and my reason for wearing women's jeans. The air cleared. The Davidsons were gracious. We all laughed, talking about wildlife and Alaska. They asked about Kevin, who had by this point decided to pursue graduate school. He would study moose and investigate whether mothers influenced the vulnerability of their calves to predators. This would help move the project toward a better understanding of learning and culture. We'd also be moving to The Hilton, as the Davidsons had graciously allowed us to use it until the project ended.

NOBEL LAUREATE KONRAD Lorenz once asserted that animals do not have culture. That was well before the world learned that Japanese macaques washed potatoes in the ocean and subsequently passed the trait to the next generation. It was well before Jane Goodall described how tool use in adult chimpanzees may be passed to young ones. And it predated the novel experiments of Dorothy Cheney and Robert Seyfarth, who found that vervet monkeys used different alarm calls to distinguish snakes from eagles. Importantly, the young of these African monkeys learned to understand the meaning of these calls. Hence, it comes as no surprise that biologists, anthropologists, and psychologists, even behavioral ecologists and philosophers, focus on and challenge the existence of culture in animals.

Such pursuits existed long before the writings of Charles Darwin and Alfred Russell Wallace. However, it has been only since the 1960s that questions have concentrated on animals in wild populations. How fast do animals acquire information about predators? Do cultures of fear exist, and if so, for how long? Do young imitate the behavior of their parents or that of group members? If sociality promotes learning, then how do solitary species like moose acquire information—merely through trial and error? Indeed, what is culture?

Among people, variation occurs geographically. American styles and traditions differ from those of the Middle East. Even within America, no one would claim that citified New Yorkers and rural Iowans are the same. Describing differences, however, and understanding why they occur are entirely different matters. Growing up in urban settings necessitates skill sets different from milking cows. Chance factors also affect how behavior develops. That only some, but not all, macaques learned to wash potatoes is no different than some cowboys who learn to play horseshoe or football.

In its most general form, culture is simply variation in behavior generated by learning and maintained within different societies. Of the commonly accepted mechanisms by which information is passed between generations, culture is but one. Genetic inheritance is another. Some purists argue that randomness is a third, but this tends to be dismissed when other factors are more readily seen playing larger roles. Beyond this general working definition of culture, nuances become highly relevant. Field scientists have generally adopted a more laissez-faire approach, while behaviorists and pure experimentalists have been stricter, manipulating conditions in the lab as well as debating semantics and issues fiercely.

In wild populations, it is far easier to identify behavioral patterns and changes over time than to understand precise mechanisms that produce change. Critics have been quick to point out that either genes or learned responses may vary with environments, so without a true experiment, one cannot claim that the behaviors observed may be attributed to anything in particular. Indeed, the patterns observed in wild nature and underlying mechanisms may at times be inextricably linked. The question of nature versus nurture has grown passé.

Because individuals vary in their behavior and propensity to reproduce, knowing individuals will help to understand processes that affect populations. Regardless of where they live, individuals face specific challenges in meeting the demands of their specific environments. If they are to promote their genes, they must produce offspring that survive and continue to reproduce. Those individuals who leave a greater proportion of offspring will be represented to a greater extent in later generations than those who do not. The suite of properties that contributes to an organism's survival and reproduction are adaptations.

Not all species' attributes are adaptations of course, for traits may persist that were useful in the past but no longer fulfill that role. A species' ancestors may have lived under conditions that differ from current ones. Some traits may simply persist because they have not been eliminated by natural selection.

My interest in culture and fear has focused on how the maintenance, loss, and reintroduction of big carnivores affect their prey at scales writ large and small. Some argue that had I been a real scientist interested in serious answers, I would have performed true experiments. By that, they meant predator removal—meaning "kill them"—so I would have some predator-free areas and others where they should be introduced. In some sense, they were correct about the design of research, that comparisons involving sites with predators absent and present were critical. But, life is never so simple, and ethical questions abound, as they should.

Given that large carnivores are in global decline, neither my colleagues nor I seriously considered killing carnivores as has been done with coyotes, skunks, cougars, wolves, and bears for centuries. On an annual basis, nearly three hundred thousand carnivores are killed in the United States alone. Other options had greater appeal.

Already, I knew moose in the Tetons had lived without dangerous predators for about sixty years. It was only in the 1990s that grizzly bears began recolonizing former habitats south of Yellowstone and wolves were reintroduced. By contrast, grizzly bears and wolves were never driven to extinction in Alaska. Perhaps my assumption that these carnivores were the primary force that drove possible differences in moose antipredator behavior was on weak ground. Moose from different geographic areas might vary in behavior for reasons totally unrelated to predators, for, as indicated above, Alaskan moose might be genetically programmed to respond whether predators were present or not. If this was the case, then even my playback experiments might produce responses to carnivores or ravens irrespective of the presence of predators and scavengers. To find out, study sites were required to replicate and compare results. Ideally, one or more Alaskan sites were needed where moose faced no predators at all.

Only one such place exists, an island below the Aleutian Range called Kalgin. Moose from Alaska's mainland had been introduced fifty years earlier, and strong oceanic currents kept them from leaving. From a scientific perspective, Kalgin was idyllic. Here were moose from Alaska, which evolved with predators but had now been predator-free for half a century. Some ten generations of calves had been isolated from seeing, hearing, or smelling a carnivore. Although harvested by humans, female moose were now protected. All I'd have to do is locate an assistant, sort out a flight, find animals, and do playbacks.

What Kalgin offered scientifically was matched by what it lacked logistically. Bounded on the west by a chain of conical volcanoes, Kalgin was 15 miles long, encircled by the frigid waters of the Cook Inlet, and scoured by hurricane-force winds. A zone of tectonic activity stretched across these Alaskan waters down to California on one side and unbroken to the west, where it connected Russia's Kamchatka Peninsula, past Mount Fuji down to Krakatau in Indonesia. This unstable sweep of active volcanoes and earthquakes forms the Pacific Rim of Fire. Forty of Alaska's one hundred volcanoes have been active in historic times. Above Kalgin Island is the steamy cauldron of Mount Illiamna, and immediately north are Mount Redoubt and Mount Spurr, both of which erupted in the early 1990s. South is Mount Augustine, which has spewed ash four times since 1935.

On good days, the unbridled beauty of these cone-shaped silhouettes reflects dramatically against skies streamed with orange and gold. A shimmering sun and clear weather are exceptions. The harshness of the island's vegetation reflects more accurately the typical weather. Thick with typical stands of spruce and muskeg, there are low-growing forests and bogs, estuaries and seaside bluffs choked with tangles. Among these is Devil's Club, aptly described in Dena'ina by Athabascan botanist Pricilla Kari as "a large shrub with . . . dense spines and prickles or thorns (that) cover the stem . . . and leaves." On Kalgin, cotton grass, tussocks, and shrubby thickets with mountain ash grow 15 feet high. Hiking trails are nonexistent, and visibility is terrible.

Unlike the openness of the Talkeetnas, Kalgin would be closed canopy. Current moose density was low, although it had once been high. Without predators, the introduced population grew rapidly because food was abundant. As the animals over-browsed the vegetation, the population crashed. Moose kneeled to feed on shoreline plants and dug for roots. Because of the silt and grit ingested, the gums of the animals were raw. Animals only four years old had teeth worn so low they looked as if they were ten.

Kalgin was also home to ravens, eagles, beavers, and a single tusked mammal. Three times the size of a moose, a walrus, affectionately called Wally, had resided on the island since the early 1980s. The population had once grown to four, but it was rumored that poachers killed Wally's herd mates. By the 1990s only Wally remained. About two hundred thousand relatives lived further north, in the Bering and Chukchi seas between Russia and Alaska. Wally was found dead in 2001, his head severed. His tusks, valued because of a resemblance to ivory—which they are not—had been removed. The senseless killing of this marine titan galvanized communities on the Kenai Peninsula and in Anchorage, where front-page articles decried the savagery.

Also on the island was a small cabin that I arranged to use. I arrived by a Beaver, a big and virtually indestructible single-engine aircraft roomy enough for a fleet of supplies. We landed on a frozen lake on a cold sunny day in mid-March. Accompanying me was Jennifer, a woman who had studied the behavior of animals in Kenya and had wintered in Colorado. She seemed to have field experience and winter savvy.

We met in Anchorage. Jenny commented that this was already the outback, as the restaurants and cuisine did not resemble those from Aspen or New York. She had wanted a good salad.

Once the plane departed Kalgin, we strapped on snowshoes and hauled gear to the unheated cabin. A network of fox tracks crossed the frozen lake. Like in much of the Aleutians, red foxes were also introduced here. To our west, volcanoes rose like soldiers, erect and capped by an emerald sun. Later, we snow shoed to the Cook Inlet where icebergs floated past. Moose tracks crossed our path. A bald eagle watched silently from a leafless cottonwood.

The next morning, my trusted backpacking stove heated water for coffee. Instantly, my mood changed. As if Pavlov's dog, conditioned by the stove's sweet purring, I anticipated caffeine in my veins even before the grounds hit the filter and aromas wafted in my direction. Jenny was a good sport. The cabin was only 4°F. Her flimsy bag could not have been warm.

We hiked for moose, finding tracks in deep snow. Venus flittered in the eastern sky, clean and azure. Mount Redoubt and Spurr changed from pink hues to full brilliance in a cold horizontal sun. My body sweated under the weight of my pack, although my toes felt like pure ice. I warmed myself by recounting the ventures of Sandy MacNab and Fred Vreeland. In 1921, these pioneers hunted and mapped areas to our southwest while carrying heavier packs. They did not have the benefit of gear from North Face, Osprey, or Arc'Teryx. But they operated in summer, not March, and not with three feet of snow. I assured myself that this still would have been a cakewalk for them.

My thoughts switched from explorers to Jenny. In the closed forest, she became disoriented. Her face was cold and stamped with pangs of fear. She didn't know how to follow her snowshoe trail back to the cabin. I worried about frostbite. I wasn't sure how, or if, this was going to work. The distance we had covered was very little, and the island was large. Jenny explained that the cold was colder than she anticipated, and conditions more difficult. She said she'd be fine reading in the cabin.

I'd be on my own. The playback unit, when combined with the 12-volt battery, was too heavy for my back. Fortunately, I had brought a sled on the plane, and had packed my climbing harness, so I'd just have to place all the gear on the sled. If I was careful I could manage the 150 feet of

16-gauge speaker wire along with the amplifier and speaker. My extra clothes, spotting scope, and cook stove would fit. I'd connect my harness to the sled using a parachute cord, and pull it while plodding on snowshoes across frozen lakes and deep snow. I was relieved Jenny would stay behind. She could get water and I would try for data. My efficiency and mental health would be better off. I had grown accustomed to solitary fieldwork.

Each day I went in different directions hoping to encounter moose tracks. When fresh, I'd follow, wishing them into real animals. River channels were best for travel, because the thick vegetation often was less dense. But the tracks shifted back and forth, usually in a bewildering array through near-impenetrable alder and Devil's club, birch and dwarf willow. The one time I encountered a moose, it fled before I could find out whether it was male or female. How they ran so quickly through entanglements without poking their eyes was always baffling. I'd never seen a blind female and supposed they were protected by thick, resilient eye lids and quick reflexes.

My reflections mattered little. The snow creaked under my weight; the sound would scare every moose around. Without being stealthy, my current plan was not working. I just wasted energy, was constantly thirsty, and hungered for my cache of dried fruits, nuts, and chocolates.

Despite being out from first light and remaining until alpenglow, my luck was not good. Three cold and sun-blistering days had produced only two solitary moose. One had thick pedicels, the bony extension where antlers once attached to the skull. The second was female. Other than looking up occasionally, she was disinterested in the calls. Neither raven nor wolf calls aroused her. Even the unfamiliar grunts of a howler monkey or the whoops of spotted hyena were unworthy of attention.

A different strategy was needed to find animals. Like in the Tetons, I'd concentrate on open areas, using my scope to check at long distances. My focus would be at dawn and dusk.

The next day, I spotted a moose three miles from the cabin. She fed in thickets and gradually moved out of site. After an hour of waiting, I was shivering and shifted to a secluded area. There, I thrashed about searching for wood buried below the snow until I had enough for a fire. Although it was 10°F, my toes and fingers were numb. Whatever warmth I once had was gone as I tried to make tea. I had been careless. The water bottles that had become my sleeping partners had frozen on the sled.

While sitting by the fire, I buried myself in woes. Why am I on a frozen spit, bound by lava, doing studies that no one in the world cares about? I have no feeling in my fingers or toes, and I can't find moose. I have a field assistant who sits in a cabin eating my most treasured foods.

Self-pity stared me in the face. If I didn't climb out of this soon, my funk would grow serious. It was time to get warm and to get data. I removed a pan from my sled, filled it with snow and turned it to water over the fire. In ten minutes the tea was hot, my body warmed, and my toes had some feeling. I plowed north, then west. At 3 PM, I saw two moose, a mother and calf, three hundred yards away. I disassembled the sled and extricated the playback gear. The first calls I played were that of the raven, then running water, then wolf. I finished the entire sequence. They had not run. Occasionally they became vigilant, looking toward the hidden speaker. Although an hour passed and I was again numb, I was excited. I had data and a packed trail to follow back to the cabin.

The next day was similar, cold and long hikes. I built two fires. The first was for warmth; but, my faithful lightweight stove quit. The water filter froze, and I couldn't dislodge the last bits of ice stuck in the lines. When I tried immersing it in water warmed by the adjacent flames, the gaskets ruptured. The second fire, later in the day, brought greater rewards. It turned dreams of Mexican food to reality. I warmed tortillas on the frying pan, and melted frozen chiles and cheese. Set against snow and sky, mountains studded with steam and forests filled with deadfall, it offered warmth and quenched my hunger. Pine grosbeaks and hairy and black-backed woodpeckers broke the silence as they picked morsels from dead branches.

With more field days came more data—playbacks to a dozen different moose. These Kalgin moose were not like the ones Kevin and I studied in the Talkeetnas. Kalgin females were not fearful of dangerous predators. They did not remain vigilant or run from sounds, whether predatory or not.

Although the Kalgin animals were genetically the same as other moose from Alaska, they differed in behavior. If anything supported the idea of a culture of fear, it was the contrast between moose of Kalgin and the Talkeetnas. Both areas are in south-central Alaska and experience similar wintry conditions. Moose from either region fed at high rates when exposed to red-tailed hawks or running water. Other behaviors were variable, however, suggesting that responses were not hard-wired. Talkeetna animals were more afraid of the sounds of ravens and wolves, ceasing to feed and preparing to flee. Kalgin animals did no such thing. Predators were one thing that varied between Kalgin and the Talkeetnas.

ON ISLANDS FAR beyond Alaska, evidence was accumulating that predation shaped behavior in natural communities. On the Galapagos Islands, finches, hawks, and lizards remain unafraid of people. Not all are totally fearless, however, since finches respond to hawks and owls but not to humans. Marine iguanas are exceptionally tame, to the point that humans can approach to within six feet and then the lizard ambles slowly away. After being caught, the same animals show heightened stress hormones to humans, but not the associated response of rapid flight. So, although a slow level of learning is reflected in their physiology, there was no marked change in behavior. Warm-blooded species distinguish more strongly and rapidly between familiar and unfamiliar perils.

In 1930, Charles Sheldon felt that squirrels in central Alaska learned about danger and modified their behavior. In Denali, he wrote, "There were six red squirrels . . . I had intended to observe through winter. . . . They had discovered my cache and had become so destructive it was necessary to kill all but one. This one was driven away so often that he finally became wary." Sheldon's last squirrel had learned predator avoidance.

Some of the most convincing evidence on the effects of predation on prey behavior stems from work with a different species of squirrel. In the foothills of Alaska's Brooks Range, arctic ground squirrels are food for grizzly bears and raptors. Dick Coss from the University of California, Davis, pieced together information about the length of time required before antipredator behavior disappears. In this frigid region, the cold-adapted squirrels live without snakes and have done so for more than three million years. Their ancestors had lived with vipers for thousands of generations before. When Coss exposed captive arctic ground squirrels to snakes they exhibited no fear. Such naiveté currently has no serious consequence, since wild arctic ground squirrels no longer encounter snakes. Other ground squirrels that still live with snakes are considerably more cautious.

Such examples illustrate that the behavior of individuals may be shaped by experience with predators, some immediate and some from the deep past. This will be no surprise to anyone with pets. Dogs and cats learn quickly when disciplined. Sheldon's surviving red squirrel also modified its behavior. But understanding how behavior is transcended from individuals into populations or species traits is complex and not readily studied in many mammals. Long-term evolutionary dynamics operating on scales that span thousands of years, such as arctic ground squirrels, Galapagos finches, or marine iguanas, can destine species to a life of naiveté. Of more immediate concern to people living off the land or tucked away in distant cities is not only whether naive prey adjust when big, dangerous predators return but how quickly.

TO MAKE A case for culture, behavioral variation must exist. If not, populations cannot move down different pathways, and all will be similar. The mere existence of individual variation in behavior, however, does not constitute culture, as information must be transmitted from generation to generation. Behavior and individual personalities are of course influenced by many processes, in part because of experience due to local circumstances, the social environment, and chance. The contribution of individuals often drives differences between populations and this offers the basis for understanding how behavior develops at a fine scale.

At a coarse level, however, behavior is also shared by common descent. For instance, people smile when they are happy; this is a trait shared by all humans, regardless of the specifics of their child-rearing, race, or geography. Smiling and laughter are species-typical behaviors, but the frequency of expression in a population will vary locally because of both internal and outside pressures. Famine or war as well as the threat of gang violence, predators, or access to amenities might govern the propensity to smile. So too might the social or physical environment. People on vacation are more likely to laugh or smile than those facing one more day of grind at an unpleasant office.

Traits shared through descent also modulate behavior in nonhumans. Populations of the same species inherit qualities from common ancestors. Just as all humans are *Homo sapiens*, all moose are *Alces alces*, and they share similar life histories and body size traits. Although Alaskan moose are larger in body size than their southern cousins, the differences are genetic in origin. Irrespective of these differences, however, males of both subspecies are antlered, urinate in ground pits to induce estrus in females during the rut, and fight for cows. All females investigate males, trying to ascertain which bulls might have the best genes.

Nuanced behavior and variation also characterizes different groups and populations as surely as they do species. Habitat, weather, predators, and other externalities all dictate specific strategies to cope, both to survive and to reproduce. Some individuals may be bold, others shy. Some might be more wary of potential predators; others may ignore them until the threat of death is in their face, on their nostrils, or gripping at their heels or throat. Individual experiences are steeped in trial-and-error learning, and the raw materials of one environment may differ from another.

So far, the case for behavioral variation among moose stemmed from playback experiments using cues to represent predators, and contrasting responses of individuals between populations. With large ears, moose detect the faintest of noises in the forest. Just as it is easier to sneak up on dogs on windy days, the same is true of moose. They are handicapped, because they cannot localize the source or direction of sounds. However, they are more apt to flee in strong wind.

Like virtually all mammals, moose may also shun danger by relying on more than a single tactic to avoid or detect predators. Beyond sound, the most obvious is a strategy also employed by humans: keeping clear of areas with bears. The indigenous Gwich'in reduce the chance of dangerous interactions with bears by camping on small islands where encounters are fewer. Female moose from Alaska and Canada do similarly when about to give birth. Even in Lake Superior, moose swim to islets where black bears or wolves are less likely to be. The use of island refuges may be a behavioral trait of pregnant females that varies little among populations because even five to ten percent of the predator-naive Teton moose also use islands as birth sites.

Hazardous encounters are avoided in ways other than choosing specific habitats and listening for predators. Eyes detect motion and minds distill forms to distinguish raven and bear from moose or elk. Smell also is critical to the antipredator arsenal. Noses with thick vascular tissue detect the wafting of a few odiferous molecules.

My challenge was to sort out how to run an experiment using the odors of possible predators, just as I had done for sound. The choices had limitations. If the dung of bears was just plopped in a field and awaited a passing moose, I might wait an hour or a day, a week or a month. What if the temperature varied, or it was windy, or if a moose moved only to within three hundred yards? Would it receive a full sensory blast, and could I repeat this on another animal? The prospects of a controlled situation were not promising.

Next I considered shooting the smelly bear nuggets toward moose with a slingshot or gun. But slingshots were not wide enough to propel sufficient quantities of juicy dung, and any sort of gun would not be allowed in national parks. Bows, on the other hand, were a possibility because of their utility in shooting more than arrows, something that British explorer Wilfred Theisger described fifty years earlier. Hunters in the arid high mountains of Chitral (Pakistan) used them; they were "about two feet long with a double string held apart at one end by a short stick; the strings were joined in the middle by a small leather pouch and the bow was used, like a catapult, to shoot stones." My inaccuracy as an archer unfortunately discouraged this option.

There was yet another alternative. I had played some ball in high school and college. I had also learned in my past studies in the Badlands of South Dakota that bison could be testy. My arm functioned just fine in keeping dangerous mothers away by pelting them with rocks, sometimes at forty yards. Although I was no Roger Clemens, I could throw strikes. Since the sound playbacks were conducted during spring, I decided to try pitching snowballs. The idea was to surgically place the morsels near foraging animals, and the distance between the odors and forager would then be measured using the length of an animal's body.

Elsewhere in the boreal north, evidence suggested prey species were highly sensitive to odors. Snowshoe hares, large, beautiful members of the rabbit family with big ears, live in areas where they must rely on more than sound to escape predators. As prey for raptors and a suite of carnivores that include coyotes, lynx, and wolverines, the hares have to be good at selecting the right habitats, knowing when to move, and sorting out mating relationships, all without being eaten. Hares are adept not only at distinguishing wolverine odors from non-harmful ones but also at avoiding areas with wolverine urine. I was keen to discover whether moose have similar chemical sensing abilities, especially given the crudeness of the system I was to use to deliver odors.

My non-technical approach had scientific promise. A blank snowball would contain only odor from my hand. I could mix the snowballs with the same amount of grizzly bear dung, and then do tosses. That way, moose responses to the added bear odor, scents beyond that of my hand-crafted snowballs could be assessed. Similar presentations could vary by substituting tiger feces or wolf or coyote urine. Other cues would include rotten potatoes, simply to determine whether predator odors differed from something novel that was not a predator. To disentangle whether responses to all urine differed from that of wolves or coyotes, I'd mix snowballs with human pee. Experiments would be done with temperatures just above freezing, and on windless days. All feces and snow would be wrapped in biodegradable toilet paper sealed with two rubber bands.

The samples came from zoos far and wide. The dung of black and grizzly bears and tigers were from facilities in Anchorage, St. Louis, and Montana. Wolf and coyote urine came from labs in Idaho, Wyoming, and Utah. All the samples were stored similarly and came from animals fed similar diets. Just as human digestion is sensitive to dietary content, what animals put into their gut affects what exits. The resulting constituency and odors had to be relatively constant.

One of the first experiments came in Denali National Park. Kevin was with me. We had just encountered a moose, and prepared to alter her chemical environment. As in our prior sound playbacks, baseline data were gathered first. Snowballs were next. After packing it tightly, I stretched my arm, preparing for a toss. The snowball had now been stamped by the smell of my hand. I launched the morsel. The whoosh caught the animal's attention and, when the snowball plopped adjacently, she looked over. The cow then continued her feeding. This snowball had virtually no effect.

After several minutes, we readied ourselves for the next treatment, the scent of human urine. Earlier I relieved my bladder, so I had nothing left. I handed Kevin the next snowball.

"Pee on this, now."

"What?"

"Drop your pants, hold the snowball, and pee on it—quickly! I'm serious. We need this to smell like a human. C'mon."

Kevin followed the orders perfectly. Soon the urine-filled and wadded ball soared. It, too, landed near the first. The moose lifted her head, extended her nose, and flicked it into the air, absorbing the finer nuance of whatever Kevin had recently passed through his system. Then, she returned to serious activity—her lignified breakfast of twigs.

THE POPOURRI OF odors did little to confuse the moose of Denali or the Talkeetnas. They ignored snowballs scented with human pee, but not those delivered with wolf or coyote urine. Apparently, it was too dangerous to hang out and smell what was in the area. Bear feces also produced strong avoidance. When it came to predator savvy, it was no surprise that Alaskan moose trumped those from Wyoming.

chapter 4

Emissaries of a Dying Epoch

Once we were happy in our own country and we were seldom hungry, for the two-leggeds and the four-leggeds live together like relatives, and there was plenty for them and for us. But the Waischus [whites] came, and they made little islands for us and other little islands for the four leggeds and always these islands are becoming smaller.

BLACK ELK (ca. 1875)

THE MIXED FORESTS and open swales of New York were once home to wolves and cougars, bison and elk. From Pennsylvania to Georgia—in fact, throughout America's original 13 colonies—the story was much the same. The hardwoods of the Atlantic harbored remarkable diversity. Woodland caribou were in Maine. Vermont had moose. Newfoundland had a flightless bird, known first as "pingouins"; in today's vernacular, the great auks.

Persecution began early, the first recorded death was in 1497. By 1844, auks were extinct. The last egg of the entire species was squashed under the boots of Ketilsson Brandsson, just off the Icelandic coastline. For passenger pigeons, the end differed in detail, not consequence. Martha—the last of her kind—died in 1914, alone at the Cincinnati Zoo. For Bachman's warblers, the finale was 1962 at the time of its last sighting in South Carolina.

In the two centuries since Lewis and Clarks' epic journey to the Pacific Northwest, changes in the United States have been rapid. The human population has swelled from six to more than three hundred million. The country's geographical center shifted from Maryland to Missouri. Ninety-nine percent of California's wetlands have been annihilated. The great plains are now the great wheat fields. In national parks in California and Utah, more than thirty percent of the species of large mammals are gone.

Some 1,200 species—everything from plants and invertebrates to snakes and mammals—have been petitioned for protective listings for fear they, too, may disappear.

Not all has been lost. In the 1920s, fewer than a thousand white-tailed deer lived in Kentucky; they now number seven hundred thousand, an abundance not universally celebrated. More than 1½ million deer die annually on US highways, about four hundred a day, events which also take a toll on human life and cost much money.

Other mammals have returned. Jaguars are occasionally sighted along desert borders with Mexico, and black-footed ferrets were put back in grasslands from Arizona to Montana. But one of the world's most heralded conservation successes are North American bison. Once numbering more than thirty million, bison were reduced by 99.9 percent. By the late 1800s, fewer than a thousand remained.

It would be less than ten years after George Armstrong Custer's fall at the Little Bighorn when an easterner named William Hornaday led an 1886 expedition to Montana's rugged Missouri-Yellowstone Divide. He managed to rope a few bison calves and bring them to the Atlantic Coast, where they eventually were displayed at the Smithsonian. By 1906, plans were underway to reintroduce bison to their historic prairie habitats. Hornaday summed the concerns, many similar to those faced by managers of other rare species today:

> There are many questions which should at once be considered. . . . Is it safe to assume that bison can be preserved for the next 500 years? . . . Is it possible to secure permanency in the maintenance of buffalo herds not owned by the states, or the national government? . . . Bison herds should be established in widely separated localities.

The next year, bison were sent west, the first to the Wichita Reserve in Oklahoma. Others followed—Montana in 1908, Nebraska and South Dakota in 1913. With wolves gone and human slaughter a behavior of the past, bison prospered. Every state has some, and a half million now roam North America.

Free-ranging they are not. Only five unfenced populations exist south of Canada—one each in Yellowstone and the Tetons, one on a non-native range on California's Santa Catalina Island, and another two in Utah. One is confined to Antelope Island in the Great Salt Lake, the other the Henry Mountains, a range named for Smithsonian's first secretary, Joseph Henry. Utah's, along with Alaska's, current bison share one important trait—neither existed a hundred years ago on ranges considered native. They also differ in another way. The northern bison still face wolves.

Canada's remote Northwest Territories, the Yukon, and northern Alberta have native predators. So does Alaska, although its four bison populations were introduced from the contiguous US. In Europe, the wisent, another species of bison, exists with wolves but only at Bialowieza, a small park in Poland. Herds from Belarus, Russia, and the Ukraine have no predators.

As early as 1875, Joel Allen hinted of a serious relationship between bison and prairie wolves: "Formerly [wolves] everywhere harassed the buffalo, destroying many of the young, and even worrying and finally killing and devouring the aged, the feeble, and the wounded."

WITH WINTER SNOW almost gone, bison herds became restless. Bulls burgeoning with testosterone played, jostled, and ran. Lacking male flamboyance, females wasted few calories. Their focus was on reproduction, not rivals nor combat. They checked other cows for signs of impending birth. Like detectives sniffing out chemical clues, these cows moved from one female to another, smelling vulvas to detect hormonal signals. When the signs became sufficiently positive, these sniffers would synchronize their gestation so that their own births coincided with the bulk of others.

Like caribou and wildebeest and a slender-horned antelope of Tibet called chiru, bison give birth during a single period of the year. The harmonization of births is considered one of the best ways to reduce predation. The logic is deceptively simple. By clumping in the same area and by synchronizing their births, mothers reduce the per capita risk on their young. Once predators have dined on a surfeit of generally defenseless babies, satiation is the expected result. Exceptions occur. As with elk and caribou calves, satiation is never a guarantee, as witnessed by surplus killing (although it has never been reported for bison).

Birth synchrony is not restricted solely to ungulates of the open plains, tundra, and prairies. Moose—while asocial and forest-living—also give birth in rapid succession. Some ninety percent occur in less than nine days, a synchrony that characterizes populations throughout North America. Like with other species, a debate persists about whether this behavior is designed to swamp predators or is simply a consequence of the spring's flush of new plants, nutrition needed by mothers to produce the nutritious milk for their neonates.

Regardless of why births are timed so tightly, bison mothers are anything but milquetoasts merely awaiting the arrival of predators to dine on their babies. They are big and dangerous and aggressive. Mothers also rely on strategies other than direct confrontation of predators. They move to the outskirts of groups to bear young, sometimes selecting broken terrain with ridges and ravines where concealment occurs for up to three days. Food intake is also reduced as the vigil to detect danger increases.

From a distant hill, I watched as a gelatinous ball of red fur wriggled spasmodically on the ground. The wind blew cold from the north. The protective mother ate the placenta, and licked her newborn's matted, wet coat. Ten minutes later, the unsteady calf stood on its spindly legs. Six minutes passed before it followed its mother. Within a day, the calf would no longer wobble. Within three, it would be able to outrun me. This was Yellowstone in late April.

Two black wolves appeared. Three newborns were within the herd. I anticipated flying hoofs and trails of dust. Other than bleats and grunts, there was little action. The cows were agitated, not aggressive. All bison in the nursery group were vigilant. Within ten minutes, there was quiet. The wolves were just testing.

To learn more about bison interactions with wolves, I'd have to rely on three geographical areas—those with wolves, those without, and those where bison were once predator-free but now experienced predation. Options of the first sort were limited to northern Canada. The Yukon and Alberta each had about four to five thousand wolves; the Northwest Territories some ten thousand. Areas without wolves were plentiful south of Canada. Only Yellowstone and the Tetons offered the third scenario—bison that had been wolf-free but were so no longer.

While Canadian wolves had never been in serious jeopardy, this was not true of their bison, whether deep in the boreal forest or out on the northern prairies. First noted in 1772, the distinctive wood bison of the far north are darker, taller, and have a more square hump than bison from the plains. In 1890, a Canadian internal report written by W. Ogilvie described them as "nearly a thing of the past. A few still remain scattered . . . [and] could some means be devised to protect them for several years, they would probably soon multiply, and become a source of food supply and revenue to the natives." In 1893, the Canadian parliament passed the first legal protection, and in 1922 a 17,000 square-mile chunk of northern wilds was set aside as Wood Buffalo National Park, population 500.

Soon after its formal dedication, a plan to bolster the population was crafted. Captive animals would be shipped north. Canadian and American scientists decried the idea, arguing that the mating between the "plains" and the "wood" bison would erode genetically important traits. In 1925,

Harvard zoologist Thomas Barbour called the decision "one of the most tragic examples of bureaucratic stupidity in all history."

Between 1925 and 1928, a total of 6,673 plains bison were shipped northward, a translocation that still reigns as the world's largest. The animals first went by train, then barge—some 650 miles—including navigation down the Athabasca and Slave rivers. The survivors of the poorly planned journey did mix with wood bison. Within ten years the population had grown with estimates in excess of twelve thousand. Tuberculosis was subsequently discovered, and the population crashed. Pure wood bison were gone, the tragic end of a once glorious forest giant.

Then, in 1959, encouragement emanated from a distant part of the vast park. An isolated group resembling original wood bison was found. Within five years, a program was in place, and 18 of the bison were transplanted two hundred miles north. By 2001, about two thousand wood bison lived along the northwestern flanks of the Great Slave Lake. Wolves, once rare in that region, soon discovered the available meat and became year-round residents. However, subsequent analyses indicated that these wood bison were not genetically pure. They, too, were hybrids, containing the DNA of plains bison.

Like wolves, human hunters have coexisted for centuries with bison. For 5,500 years, indigenous peoples drove them over cliffs on the prairies of southern Alberta. The herds were funneled for at least three miles across fields three hundred yards wide, then driven over cliffs ten yards high. The spectacular death assemblage, known as Heads-Smashed-In Buffalo Jump, is now a protected area. To the south, bison were also hunted and revered. Bison masks, shields, and head-dresses were worn in celebration by tribes of the Gros Ventre, scrapers and paintbrushes used by the Lakota, and both the Comanche and Kiowa adorned hides with drawings. The Shoshone even depicted bison on elk hides.

In 1841, the relationship between Plains Indians and bison may even have protected other ungulates from predators. George Catlin noted their unwariness: "The buffalo herds, which graze in almost countless numbers on these beautiful prairies, afford [Indians] an abundance of meat; and so much is it preferred to all other, that the deer, the elk, and the antelope sport upon the prairies in herds in the greatest security, as the Indians seldom kill them." In 1865, Canadian John McDougall witnessed on the northern plains "more buffalo than I ever dreamed of before. The woods and plains are full. . . . [W]e came to a large round plain, perhaps 10 miles across, and as I sat on the summit of a knoll looking. . . . [I]t did not seem possible to pack another buffalo into the space."

Neither bison nor North American Indians were to survive in the wild much longer. By 1875, bison populations had crashed from millions to a few thousand, their destruction a blight for American humanity. Not only was an innocent species viciously slaughtered for hide and tongue, but people and their cultures collapsed. "Principal Indian tribes on the plains, being deprived of their annual supply of food for the winter, and only receiving short rations on the reservations, [were] driven on the warpath," so commented Colonel Richard Irving Dodge, of the US Army, in 1875. In 1893, a group of Lakota women and children were massacred by US soldiers in the Badlands at a place called Wounded Knee. The rapacious had struck; four-footed wolves they were not.

WILSHIRE BOULEVARD IS lined with palm trees. It runs from downtown Los Angeles westward to the Pacific. Along the route are skyscrapers, fancy cafes, and the University of California, Los Angeles. There are Pizza Huts and McDonalds, dingy bars and organic grocers, Vietnamese restaurants and Mexican taquerias. The street is a slice of modern America.

When geologist J. Whitney described the LA Basin in 1865, he found dirt roads and a site with "a very large amount of hardened asphaltum, mixed with sand and the bones of cattle and birds." The tar pits of Rancho La Brea trapped extinct bison, camels, and saber-toothed cats. A museum was built on the site in 1913, around which the City of Angels prospered. By the 1960s, when warm coastal breezes from Santa Monica Bay blew the smog eastward, it piled up against inland mountains—places called Hollywood, Pasadena, San Gabriel, and San Bernadino. The valleys below grew fruit trees and crops.

I grew up near Wilshire and spent summers at the beach and at Dodger Stadium. In my teens, there was body surfing, baseball, and dreams of scantily-clad women. The Watts riots followed, and my friends and I were beat up. This was LA, in all its good and bad.

Beyond Santa Monica were the golden beaches of Topanga, Malibu, and Zuma. My Harley took me to the Big Sur coast, to the Sierras, and to the searing deserts of Nevada. Later, a BMW motorcycle took me east and north, hiking and backpacking in Canada. My interest in the natural world varied inversely with that for my fellow humans. While in college in southern California, I worked the orchards, picking cumquats and grapes, almonds and avocados, grapefruits and persimmons.

In graduate school in Colorado, I wrote a dissertation contrasting bighorn sheep in Canada with those in deserts near Mexico. Then I moved east. The Smithsonian had offered a postgraduate fellowship in animal ecology. From there I came west again—this time to study wild horses in the Great Basin Desert of Nevada.

One day while visiting Santa Barbara, I met a stunning woman with a killer smile. Her name was Carol. We would study horses, hike in mountains and bathe in hot springs. Later we would marry. During this period, I accepted a faculty position at the University of Nevada in Reno. The hiring package enabled me to concentrate my teaching into the fall semester. During springs and summers, I would focus on research and graduate students, an arrangement that would last 16 years.

THE LOS ANGELES BASIN was once a wild place. It still is. There are floods and earthquakes, potholes and traffic, gangs and murders. There are also coyotes and cougars. Thirty years ago, condors soared above snowy mountains with bighorn sheep. Grey whales and dolphins still pass Santa Monica and Malibu. On Wilshire, the tar pits of Rancho La Brea continue to yield knowledge about the region's early inhabitants, emissaries of a past epoch.

During the late Pleistocene, pools of death—the result of asphalt oozing from alluvial sediments—trapped a fauna whose largest members are all but extinct. Almost 60 species of mammals and 140 birds have been noted, spanning the last 40,000 years. Among them are giant ground sloths and camels, some seven feet tall at the shoulders. Two species of horses are present, one ass-like, the other zebroid. Extinct bison, however, are the most frequent, the "long-horned" and the "ancient," each species larger and more substantive than our currently surviving stock.

Less common are deer, tapirs, and mastodons. Among the smallest of the grassland browsers is a dwarf pronghorn, only a foot and a half tall. Long-headed llamas, wild relatives to South America's native camels—guanacos and vicunas—are also present.

Given the abundance of herbivores, it is no surprise that the area was home to many carnivores. Where apartments, markets, and playgrounds now stand, dire wolves and two species of saber-tooths roamed. There were even coyotes. Overall, seven species of fossil cats have been recorded. The abundance of meat-eating mammals and birds is most likely explained by their attempted feeding on carcasses already mired in the asphalt bogs.

About the same time that mega-animals were living and dying in the grasslands of LA Basin, traces of a comparable fauna would be noted elsewhere. Steppe bison grazed along with the bizarre proboscis-laden saiga antelope, as well as horses, camels, and mammoths, in an uncanny assemblage known from the cave drawings at Lascaux in France. However, predators such as wolves and lions—the latter the same in appearance as those from today's Africa and Asia—stalked these prey.

One can imagine a lone steppe bison bull grazing along a creek bottom. Commotion erupts and two lions attack. The bull succumbs, just as Cape buffalo in Africa do when assailed by lions The carcass is never completely consumed, and over time it is covered by snow and then by silt, where it entombed for 36,000 years.

In 1979, a miner named Walter Roman unearthed the frozen carcass of a steppe bison in central Alaska. It was examined by a paleo-ecologist, Dale Guthrie, and named Blue Babe after Paul Bunyan's giant ox. With blood clotted under skin, canine perforations and scratch marks on the hide, and puncture wounds on the face, Dale concluded that Blue Babe had been killed by lions.

Knowing that the ancestors of modern bison were likely killed by lions meant that I would have to include the calls of large cats in addition to wolves in my playback experiments. As with moose, I'd include ravens, particularly since early reports suggested an interactive web among bison, wolves, and ravens. This ecological dynamic was noted on the Great Plains 150 years ago by James Mead:

> The reasons we put out our baits [for wolves] after sunset was an account of thousands of ravens that seemed to live with the buffalo. . . . [The ravens] would come back and pick the baits [strychnine] if put out before dark, so that instead of killing wolves, we would find we had a whole field of ravens killed. . . . The buffalo, the gray wolves, and the ravens—companions in life—mingled their bones when swift destruction overtook them.

WHAT THE GREAT Plains, the LA Basin, and Alaska offered in the late Quaternary and what America's grasslands offered ten million years ago is akin to what Africa offers today: spectacular diversity, at least in a few surviving savanna enclaves. The contrasts did not elude early scientists, especially Henry Fairfield Osborne of the New York Zoological Society, who noted in 1904:

> These [extinct] animals flourished during the period in which western America must have closely resembled the eastern and central portions of Africa at the present time. . . . [T]he resemblance between America and Africa is abundantly demonstrated by the presence of great herds of horses, of rhinoceroses. . . . of camels in great variety. . . . of small elephants, and of deer, which in adaptation to somewhat arid conditions imitated the antelopes in general structure.

Today we know these savannas differ. Seasonality and the history of the two continents are dissimilar. Nevertheless, contrasts between African and American sites are replete in the media and by conservationists if only because intact sites in America are so few. Two areas have regularly been compared to Africa's Serengeti, perhaps because some semblance of large mammals has been retained in the United States. However heuristic the contrasts, they pale when real numbers are considered.

Yellowstone is twice the size of Serengeti, but it has only 7 native ungulates, compared with Serengeti's 31. Alaska's Arctic Refuge has just 4—musk ox, Dall sheep, moose, and caribou—despite being three times as large. When reserve size is taken into account, the differences in diversity are accentuated. The East African reserve with the fewest ungulates is Mount Kenya. Just one-fifteenth the size of Yellowstone, it has twice as many large mammals. The smallest area, Ngurdoto, is less than 25 square miles yet boasts 18 species.

While the North American parks maintain relatively few species, the contrasts appear worthwhile if for no reason other than congregations of ungulates still occur in North America. But continental differences are clearly notable, and are reflected in management philosophy. Roads crisscross the Serengeti. Arctic National Wildlife Refuge has none. Some two hundred thousand Serengeti ungulates die in the snares of poachers each year, whereas legal hunting of all big game is permitted in the Arctic Refuge. Conservation efforts far beyond the Arctic Refuge, Yellowstone, and Serengeti have been substantive and underscore a richness which still remains on both continents.

WITH SNOW FALLING and the last of the semester's exams over, I put winter gear and playback equipment into my VW van. Winter road closures and pure ice meant that the short 750 miles between northern Nevada and Wyoming would be treacherous. One 125-mile stretch required 48 hours. After a five-day trip, Carol, Sonja, and I—and our two dogs—happily reassembled in our small Wyoming cabin outside the hamlet of Moose.

By February, the Tetons were under brutal assault. Clouds hid the peaks for 16 days. To enter our abode, we walked through a narrow corridor with snowy walls nearly ten feet tall. To avoid a collapsed roof, I shoveled blocks of hard packed snow from the top of our cabin. Sonja and our dog Sage jumped from the roof into the snowy foam below. When the skies finally cleared, it was as if a vacuum had sucked any remaining heat from this part of the planet.

One morning, Carol and I drove Soni to school before heading out for fieldwork. It was –37°F. The gas line froze, shutting down the car. I walked for help. With no face mask and but a slight breeze, the temperature on my skin felt like 50 below. My nose turned red, then white. Help was not around the corner, only frostbite.

Carol and Sonja remained, shivering and in serious pain. After two miles, someone picked me up. In the meantime, Carol had been innovative, thawing the gas line with isopropyl alcohol. Thereafter, our vehicles always carried heavy sleeping bags, more alcohol, a propane-powered heater, extra heavy socks and mitts, and an emergency kit.

IN EARLY APRIL the bison grew restless. After their winter's absence from Grand Teton Park, they trickled in from feed grounds to the south. First to appear were bulls. Cows and calves appeared a week or so later in groups of twenty, thirty, or fifty. The migration had begun.

Bison in the Tetons and Yellowstone were likely to be predator-free, since wolves had been removed from the landscape more than sixty years earlier. However, because neither adult nor calf survival had been monitored, it was impossible to know whether coyotes or bears were having any impact. The possibility seemed doubtful, because population increases in both parks mirrored our findings from South Dakota's Badlands, an area with no bears where better than ninety percent of the calves survived.

If bison adults or calves were not preyed upon, then—as with moose—animals from these predator-free populations should have no incentive to respond to the calls of wolves or ravens, or even the odors of wolves or bears. Since bison and moose require thirty to forty pounds of vegetation to meet their daily energy needs, wasting precious time by responding to sounds, whether familiar or not, seemed unnecessary.

Determining if this prediction was true was the first part of my program aimed at understanding bison perceptions of predators. I would first do experiments in the contiguous US before flying north to Wood Buffalo National Park in Canada. I expected bison from the Lower 48 and Canada to vary. The former should be indifferent to the sounds of wolves and ravens, but Canadian bison should be relatively astute, given predation upon them by wolves.

My playback arsenal would contain the roars of lions. I wanted to gauge sensitivity to past predators. Since all living bison share in their ancestry a strong genetic link with steppe bison and the North American lion preyed on the steppe bison, I anticipated that both US and Canadian bison should be either equally responsive to, or disinterested in, lion roars. In essence, responses to an extinct predator should be invariant, whereas modern bison should modify responses to current predators based on their recent experience with predators.

During her spring break, Sonja and I drove from the Tetons north to Yellowstone in a van filled with playback gear, snowshoes, wires, cables, sleeping bags, and food. She wondered why we could not be en route to Australia, Indonesia, or Mexico, locations to which her friends were heading. After spending her early years sleeping in the back of a Land Rover in Africa, camping was not Sonja's idea of fun. I understood her point, but this trip was to gather exciting new information on bison. I knew bribing her was a poor idea. Simply because her friends had wealthy parents did not mean that was a road we could afford to travel. Besides, she liked the idea of tricking animals into thinking that wolves might be lurking over the next hill.

The first bison group we found was four hundred yards out in a wet meadow. Granite boulders and a small patch of conifers offered us good hiding. I carried the 12-volt battery and amplifier, Sonja took the rest. Concealed, we watched the unwary herd. Sonja used the laser range-finder to estimate the distance between us and them—275 yards. The measure was critical, for if bison perceived the threat to be sufficiently close, they might respond differently than when it seemed far. By always noting the distances between the speaker and bison, it would be possible to determine statistically if a threshold existed upon which the sounds had either important or little influence. We also measured the height of vegetation, because if bison were concealed in forest or tall grass and they could not see, they might be more likely to be vigilant or flee. In this case, we gauged the vegetation to be no more than eight to ten inches high by contrasting it with how much of their legs remained in view.

Earlier in the day, a black bear had fed upon new grass only a couple of miles away. Three coyotes had been feasting on the rancid meat of an elk that starved. The wind was nil, the skies cloudy. A few snow drifts lined the meadow, and two ravens flew overhead before alighting near the bison herd.

Next in our data diary was an accounting of calves as well as the sex and number of animals in the group. Our final chores were to decide which animals to sample, and to note their position within the group. Animals on the edge might be more likely to look up than those in the center, since those on the periphery might be more sensitive to sounds or disturbance because of increased vulnerability to a predatory attack. Although we didn't know if this would be the case—and it seemed unlikely here in Yellowstone—we had to employ the same methods that would later be used in Canada.

Sonja took the nearest female, an animal at the group's edge. With greater experience, I opted for three adult females in the herd's center. Before trying the playbacks, we recorded the rate of "normal" feeding. All animals mowed down vegetation, and of our four data points that reflected feeding activities for 180 seconds, each animal received a hundred percent for the allotted sampling period.

We readied for the playbacks, choosing ravens first. Sonja's bison became vigilant after eight seconds, and remained so for another nine, returning to feeding even before the playback ceased. My three animals all fed the entire time. They ignored the invisible ravens. Next were howler monkeys. We stayed focused on the animals we had been monitoring. Sonja's animal again flinched, but fed overall for about ninety percent of the observation bout. My three females varied, from sixty-five percent foraging time to one hundred percent. We followed with wolves. No one lifted its head. Then came running water and crickets—these were controls—sounds that represented no potential danger and something that bison were familiar with, even if it was too early in the year for crickets. The responses were nil.

Our final calls would be those of lions. The roars started slowly, a slightly muffled "aou," followed by a deeper "aoh," and then another increasing in pulse and intensity, and another and another. Like the other playbacks, all acoustics were the same volume loudness and durations, fixed at 22 seconds. By the time 15 seconds had elapsed, every bison was looking toward the granite boulders that concealed us. Once the lion roars stopped, about half the animals returned to feeding. Others took longer to adjust; one animal did not return to feed for 300 seconds. None fled the area.

Why had the roars of lions triggered such a massive response? Was it the small group size? Larger groups contain more ears, eyes, and noses, so when one animal becomes alarmed it might trigger a response in others. Maybe it was the tonal qualities of the calls themselves and it had little to do with the lions per se. Or, perhaps, as some might delight in claiming, bison had historic memory of lions, something of which I remained skeptical.

Sonja and I scoured Yellowstone for more bison. We did playbacks, sometimes hiding behind or in the car, and often on the ground. Group sizes varied from a half dozen to seventy, our distance to the bison from sixty yards to six hundred, and habitats from forested to grassland. By the time we finished, we had measured the responses of more than a hundred bison. The only sound that produced excessive wariness was the roar of lions. I wondered whether this was a peculiar trait of Yellowstone bison or bison everywhere had the same reaction.

We headed south. Sonja's vacation was nearly over. That night we camped in an isolated canyon in a blustery snowstorm. The outside thermometer dropped to near zero, but inside the van it reached a balmy 68°F. We brewed hot chocolate and enjoyed a late dinner of pasta with pesto. We laughed and told silly jokes as we drifted to sleep. By morning, the wind whipped more snow as I scraped frozen condensation from every inside surface.

During the next few years, I continued my pursuit to understand the bison view of predators. I conducted more playbacks in the Tetons, Yellowstone, and some in the Badlands. My sample sizes grew. The species that bison "feared" most was lions. The howls of wolves and caws of ravens were barely acknowledged. Red-tail hawks had as negligible an effect as howler monkeys. It did not matter if the sounds were familiar or foreign. In an arena where bison were no longer prey, they were unresponsive.

Only when I returned to Yellowstone years later, after wolves had killed at least 16 bison, did bison notice their howls. Some individuals grew more nervous, although I did not know if these were mothers who lost calves or individuals that had had serious encounters with wolves. Overall, the responsiveness of bison in Yellowstone was low, but more individuals showed heightened alertness than in prior years. The interest held by Yellowstone bison in wolves was not shared by Teton animals where only a single bull had succumbed to predation. Ravens elicited no interest, but lion roars captured the attention of almost all bison. I hoped Canada held clues to this mystery.

I remembered the words of Lu Carbyn, a Canadian biologist with interests in peace and tranquility, wolves and their prey: "[bison] have not a care in the world, they lie down, and chew the cud . . . while older youngsters playfully chase . . . like the Garden of Eden. Yet, always, the peace treaty is only temporary [for as soon as wolves arrive], and soon Eden becomes Gallipoli."

part ii

The Meek and the Bold

The wolf is one of the wildest and shyest of all animals. . . . Where they have not been exposed to people for generations, wolves may act quite tame.

L. DAVID MECH (1970)

chapter 5

Subarctic Shadows

Worry gives a small thing a big shadow.
SWEDISH PROVERB

FIGURE 16. Aerial view of Wood Buffalo National Park in Alberta, Candada, and a group of about ten bison (visible in the bottom left of the photo) and their food craters.

FIGURE 17. Wolves on a bison in Wood Buffalo National Park. (Photo by L. Carbyn.)

FIGURE 18. Mike McMurray prepares to ford the Oshetna in Alaska.

FIGURE 19. Tom Roffe crossing the Snake River in Wyoming.

FIGURE 20. Female moose during the bonding phase with her newborn.

FIGURE 21. The Russian village of Terney on the Sea of Japan.

FIGURE 22. Olga with remains of a moose killed by a brown bear.

FIGURE 23. Moose (*los*) in the Kolumbe River Basin of the Russian Far East. (Photo by O. Zaumyslova.)

FIGURE 24. Cabin in the Kolumbe River Basin.

FIGURE 25. Climbing ledge used to observe moose.

FIGURE 26. Sound playback unit, wire, and speaker.

FIGURE 27. Three Inuit hunters, sled dogs, and caribou hides in Greenland. The hunters dressed in polar bear pants.

FIGURE 28. Musk ox family group, three generations.

FIGURE 29. Flying above Svalbard, ten degrees from the North Pole.

FIGURE 30. Kim gathering snow for water in a –60°F breeze on Svalbard.

THE PLACE CIRCLED on my map had been described half a century earlier by two biologists as "an intervening area of extremely inhospitable and unproductive terrain of karst[ic] topography, poorly defined drainage, and much muskeg interspersed with high sand eskers."

This was a land of water and prairie, bison and wolf. If there was a spot on the globe where bison should respond to wolves, Wood Buffalo National Park was it. Sequestered in the vastness of subarctic Canada, this Denmark-sized reserve is the planet's only region where bison remain the mainstay of wolves. Twenty years ago, the United Nations designated it a World Heritage Site. Remote, with no services and little access other than by air or water, this 17,000-square-mile wilderness still receives fewer than two thousand tourists annually. If solitude were the goal, subarctic Canada would be the place.

I needed to find the man who produced two books on the region, Lu Carbyn. Born in Namibia, skilled in boreal survival, and with a PhD in biology, Lu worked for the Canadian Wildlife Service. I hoped he would facilitate permission for my studies. The process of acquiring research permits is fairly universal: convince authorities that the scientific question is legitimate, that the investigator is credible, that the study has some value, and that no one will die. If I could do so, Lu would meet me in Alberta's largest city, Edmonton, and we would go north together.

There was yet one additional hurdle. It concerned the burden of proof. Canadian authorities required assurances that my playback experiments would not alter the distribution of wolves. The idea that I was expected to verify something in advance of study posed an interesting dilemma. I could easily envision that resident wolves would, out of curiosity, approach the location from which the howls were played, perhaps treating them as if a novel pack had invaded their territory. If this were the case, did it really mean that wolf distributions were altered? Or, if the wolves fled from the sounds, a scenario that seemed rather unlikely, would that also constitute redistribution? I had no way of answering that question, since the wolves of Wood Buffalo Park were not radio collared.

In my application, I indicated that I could not offer the desired assurances. Scientists from three US national parks—Yellowstone, the Badlands, and the Tetons—all shared similar legitimate concerns about possible effects on other species, yet they had granted permission. I included their letters of support in my request to work in Wood Buffalo. Canadian authorities asked a few more questions and then granted permission.

In addition to sound playbacks, I had also planned experiments with the scents of wolves, bears, and tigers. In the three US parks, bison had been approachable but unresponsive to the simulated predator cues. If Wood Buffalo animals cooperated, I was not going to squander an opportunity to deposit scents. I was unsure if March temperatures would be conducive, but I planned to bring along scats and urine just in case.

At the Canadian customs office, the uniformed border agent immediately honed in on my oddly-shaped packages. I explained why, unlike other visitors, I had wolf urine, tiger and bears scats, a slingshot, and an assortment of playback gear. I was less than convincing. An Agriculture Canada officer was next, and he insisted on permits. Then the queries for illegal paraphernalia began.

The agent's eyes latched onto mine. He waited, hoping for me to flicker, blink, or otherwise expose contraband. This was not good. I had not planned on spending two full days in a motel awaiting clearance for my gear, nor had Lu Carbyn. We had allocated only ten days to fly up to Wood Buffalo, find a truck, find a plane, and find bison.

The wolf and coyote urine were stored in Nalgene bottles. If the seal remained unbroken, odors would not escape. The tiger and bear dung were a different story. They'd been frozen in baggies, but after sitting at room temperature for 36 hours, they were starting to ooze. A different Canadian customs official asked what stunk and demanded to see the permits to import products from endangered species.

I took a deep breath, and explained that the tiger and bear scats had come courtesy of the Anchorage Zoo. In my politest demeanor, I pointed out that even when species are protected by the Convention on International Trade in Endangered Species, it was legal to possess by-products as long as the animal was not harmed. He was not fond of my lecture, but Lu and I made the next flight to Fort Smith, a community of 2,500 people on the Slave River, just north of the park. The melting feces followed.

From above Edmonton, the manicured landscape transitioned into a snowy mosaic filled with gently rising plateaus and frozen rivers. Fire-scarred forests, bogs, and meandering streams replaced wheat fields and roads. Marshes, shallow lakes, and vast frozen meadows of sedge appeared. The Peace-Athabasca Delta lay below.

At Fort Smith, we checked in to the Pelican Inn and rented a beat-up blue truck with a windshield filled with cracks. Our plan had four steps. First was aerial recon to locate bison. Next, we would land the plane a few miles away, strap on snowshoes, and hike until we reached the bison. The sound playbacks would follow. Finally, the plane would come back and pick us up.

The plan, although simple in theory, had possible obstacles. Bison might not be detectable from the air either because the area was so immense or they could be obscured by tree cover. The weather might turn sour, or thinly-iced rivers might prevent our crossings. Even the sound equipment might fail.

Although the National Science Foundation had approved a grant, one reviewer noted that it made more sense to record the animal sounds on CDs rather than relying on the more archaic tapes that I had been using. The suggestion had merit, and I had in fact tested the CD option earlier. At –10°F, I discovered that the electronic equipment skipped tracks. At 20 below, it failed altogether. Already I worried about the power drain on my 12-volt battery in the cold, and did not need the additional burden of electronic uncertainties. Cassettes it would be.

With the threat of an incoming storm, we adopted Plan B, attempting to find animals from the ground. Lu and I spent the first three days driving slowly some 120 to 250 miles daily on snow-packed dirt. Bison were nowhere to be found. Spruce grouse, northern goshawks, and willow ptarmigan were exciting substitutes for megafauna. Our path crossed the tracks of five wolves in early morning. At dusk, a feline shadow flitted from tall grass and stepped into the open. The lynx eyed us for thirty seconds before disappearing into the chilled landscape.

At a cabin at road's end, the butchered carcass of a frozen moose greeted us. Dinner. Nearby was a power saw. This was the home of Archie, a Cree-Chippewa who had hunted these forests and fished their waters most of his life. Lu wasted no time in querying his friend. Archie had just returned from checking trap lines on his snowmobile. Bison had not been around for weeks. Other species were plentiful; ten marten, two lynx, and three moose had been harvested.

The next day, the weather cleared. We were ready to fly but lacked wings. At Loon Air, the plane I had booked for three full days was no longer available. We moaned to the attendant who indicated the plane could be available for a total of four hours. Thereafter, extensive service was necessary.

We were screwed—no plane, no bison. Lu knew of another pilot, but we couldn't find him. So we climbed back into the heap of a truck, drove almost three hundred miles more in the hope of finding bison. Our dismal luck remained. Archie was correct—the bison had gone south.

Back at Pelican Inn, we talked of life, careers, and Africa. As the conversation shifted from data to disillusionment, my frustration over the lack of bison and aerial support grew. This was the challenge of fieldwork, which often ran as smoothly as an old jalopy bouncing down a corrugated road. Rather than wallow in sorrow, I strapped on a headlamp and headed out for an evening jog. Stumbling across a local cemetery, I examined names and dates while absorbing the inland delta's rich indigenous history. Back at the motel I cooked our dinner on my small stove, flavoring rice with savory spices and garlic.

Dawn broke clear and cold, about 0°F. Before dripping coffee, I looked down the street, still silent and dim. Garbage bags were everywhere. Who would have made such a mess?

All the trash came from a single vehicle—a blue truck with a broken windshield. Ours. Two ravens were in the back, pecking at something frozen. A third flew, weighed down by a debris-filled beak—a morsel of tiger shit. The ravens had eviscerated and scattered the trash bags. Bear poop was the preferred delicacy; there was none left. Our opportunities for the scent experiments had vanished, though the cold temperatures would have dampened any perfunctory smells anyway.

We again checked on the availability of a rental plane. Our luck had changed; a large Beaver sat already mounted on skis and was available. Soon we were off. A small group of bison stood near patches of leafless aspen, but they ran even though we were 1,500 feet above. We found more animals 35 miles south. Such distance means little to bison. Lu once back-tracked 90 animals that had run after being attacked by wolves. Nothing remarkable there, except the bison covered 50 miles in snow during just 12 hours.

A group of thirty appeared in the open. We banked hard and landed in a frozen glade. We guessed the bison were only two to three miles away; tall stands of pine and spruce separated us. With snowshoes strapped on and playback gear on sleds, we were ready to hike. I added medical supplies, food, and sleeping bags to our duffel. If the plane did not return, we'd have supplies to survive.

Our progress was slow. After half a mile, we did not know which direction to take. The animals had vanished, and all we could see was the flat boreal forest. To expand our survey we left equipment, split up, and broke track alone, agreeing to meet back at the cache in an hour. With each step I plunged through crusted snow. Although my fingers were numb, sweat dripped from my brow. At 9:30 AM, a horizontal sun emerged, its golden rays softly touching treetops. The water in my pack had already frozen. By 10:40, Lu and I reunited, each frustrated at our lack of success in finding bison. He saw a red fox. I scared a grouse. Our search took us in a third direction. We crossed a frozen creek.

By noon, we emerged onto a prairie rimmed with willow, horsetails, and rushes. Deep pock-marked craters filled the snow, areas where bison had swept it aside using their powerful necks and heads. Underneath was grass and sedge. Soon we encountered dung, some not yet frozen. Fresh tracks followed. Our quarry was close.

A group came into view. Sixty seconds later, Lu detected another group, then a third. I checked the distance of the first herd—five hundred yards. I pulled a tarp, setting the battery gently on it, followed by the speaker. Lu unthreaded the spool of wire that connected the speaker to the amplifier. I grabbed for the satchel with audio tapes. Lu was talking, but I couldn't hear a word he uttered. As if in a nightmare, his words were slow, muted, garbled.

Panic set in—where are the tapes? I'm dead. With no tapes, this is all for naught. We had come hundreds of miles north from Edmonton, rented a truck, driven for days, and been dropped by plane. The tapes were missing. My heart and my temples pounded.

Madly, I pulled more and more gear off the sled, flinging it in haste. Finally my hand hit something familiar—the padded satchel containing the magical sounds of ravens, wolves, tigers, howlers, and more. I relaxed. Lu's voice returned, asking for more instructions about how to attach the cables.

We played the spectrum of sounds. The bison barely blinked. A couple looked up momentarily when ravens cawed. Two other females did similarly, one for wolves, the other for monkeys. Overt responses these were not. Even hyena whoops drew little attention.

I tried the lion roars. Within 18 seconds, all bison stared. Animals on the edge of the group moved inward, as if group defense was about to occur or animals would flee en masse. Within three minutes, most resumed feeding, although a few waited six minutes or more. None fled. We hiked more and repeated playbacks to different groups.

The plane arrived a few hours later. During takeoff the skis wedged into our virgin, snow-packed runway. Immediately we lost speed. Willow thickets and dogwood blocked our path. It was too late to abort. We accelerated, clipping their tops but gaining altitude, once again rising in search of bison.

Our efforts proved successful. Distances between animals and us for playbacks varied from 600 yards to only 75. The habitats we selected were diverse, ranging from prairie to forest, and with little visibility to wide openness. We sampled nearly two hundred animals. Irrespective of our distance and cover, the behavior was nearly the same—bison were unresponsive. My measures for fear—vigilance, grouping, fleeing—were barely detectable except when lions roared. Even then, responses were mild. Were bison savvy or fearless?

THE WELL KNOWN phrase "Nature, red in tooth and claw," written more than a century ago by British poet Alfred, Lord Tennyson, is what most people think of when predation comes to mind—overt, brutal, and bloody. The act of killing existed long before the first mammals appeared more than 230 million years ago.

In both the past and present, predation and predators enamor. Paleolithic drawings of lions and bears line cave walls in France. Egyptians kept pet cheetahs. In 1790, James Bruce wrote of hyenas "[they] had given occasion to so much confusion and equivocation. . . . [I]t began very early among the ancients." But it was Charles Darwin who thought about process and pattern, individuals and variation, noting in 1859: "No one supposes that all individuals of the same species are cast in the same actual mold. These individual differences are of the highest importance for us, for they are often inherited . . . and they thus afford materials for natural selection to act on."

Darwin was prescient—some ingredients of predator avoidance, such as vigilance, grouping, or fleeing, have an innate component. Others are more labile. The rate at which social behavior and associated skills develop is modified by peer groups, parents, and early experiences. Even food, toxins, and past sexual history shape current behaviors. Antipredator efforts come in many forms, not all associated specifically with avoiding dangerous carnivores.

Prey also modify their behavior in response to other threats to their fitness. Insects and parasites are two such dangers. In 1860, the year after Darwin published *On the Origin of Species,* George Armstrong Custer commented on hordes of tormenting mosquitoes on the northern prairies, "to such an extent do these pests . . . exist that to our thinly coated animals, such as the horse and the mule, grazing is almost an impossibility, while the buffalo, with his huge shaggy coat can browse undisturbed."

Any doubt that insects can shape the behavior or immediate habitat preference of at least one large-bodied mammal was put to rest in 1866. J. K. Lord described the torment:

> Traveling in Oregon, one constantly finds oneself on the banks of a wide glassy lake; gazing over its un-rippled surface, the eye suddenly rests on what to the inexperienced in hunter's craft, appears to be small clumps of twisted branches, or dead and leafless tree-tops, the trunks of which are hidden in the water; but the Indian and trapper discerns in a second that the apparent branches are the antlers of a herd of Wapiti that has been driven into the water by flies.

Just as individual moose attempt to increase their odds of survival by giving birth away from predators on islands, elk, bison, caribou and other species also adopt similar strategies to minimize harassment by insects. Subtle behaviors underscore life-and-death struggles that go unnoticed by the lay public and scientists alike. Some individuals use home ranges that have higher quality food. Others migrate to high altitudes or move hundreds of miles even though other members of the same population are sedentary. It is this fine variability to which Darwin referred:

Often it is not "nature, red, in tooth and claw," or the dramatic battles of antlered or horned rivals, that determines who lives and who dies. Instead, as noted by Harvard zoologist Ernst Mayr, "The struggle for existence . . . rarely takes the form of actual combat. Ordinarily it is simply competition for resources in limited supply."

THE COLLECTIVE LACK of response by bison to predator cues was unnerving. The northern bison had predators, while those far to the south in the Badlands and Yellowstone regions had none or only recent experience with wolves. I did not understand why wood bison did not bother to worry about wolves. And, I had a difficult time reconciling why two species, bison and moose—both experiencing similar predators—responded so differently.

Each was preyed upon by wolves. The intensity of predation was not trivial in Alaska or northern Canada, where population growth in each species was limited. Additionally, young were primary targets, often being killed at disproportionate rates relative to their frequency in the population. If anything, both moose and bison mothers should have responded to cues that signified the presence of serious carnivores.

The similarity in responses by bison to wolf howls was mirrored by their feeding rates. Animals from Wood Buffalo were essentially nonvigilant, a pattern virtually identical to those of bison from Yellowstone, the Badlands, and the Tetons. Bison were feeding machines, constantly satiating voracious appetites rather than scanning for predators. By contrast, Alaskan moose from Denali and the Talkeetnas responded to the sounds, but those unfamiliar with predators showed few reactions.

In other ungulates, especially those from Africa, relationships between vigilance and predators have been clear—the greater the magnitude of predation, the more vigilant and flighty the prey. Thompson's gazelles in the Serengeti eye their major predators—cheetahs—at distances greater than those when watching hyenas or lions. Wildebeest, of which fewer are killed by cheetahs, are less vigilant to them than to predators with greater hunting prowess—lions and spotted hyenas. Where hunted by humans, both gazelles and wildebeest become more vigilant.

Did bison distinguish among different types of predators? The evidence was teasing and inconclusive. Drawings by artist George Catlin in the 1830s showed Indians disguised in wolf hides stalking bison on their arms and legs. Given that Indians approached bison in this manner, rather than exposed, it seems reasonable to presume that wolf camouflage enabled closer approaches. Just as wildebeest tolerate cheetahs in closer proximity than lions, wolves may have been able to approach bison more closely than humans. If so, then bison must have discriminated among possible predators.

Why wolf-savvy bison from the Canadian north didn't even lift their heads to wolf playbacks was a different matter. Perhaps bison just never respond to wolves. Hearing a distant wolf howl may indicate little about their readiness to hunt. After all, noisy predators that reveal their presence may have few opportunities to hunt successfully. Or, it may be that bison have learned that howling wolves are not hungry wolves. Nonetheless, nineteenth-century naturalists reported coassociations between wolves and bison on the prairies, a scenario not dissimilar from Lu Carbyn's descriptions of bison and wolves in northern Canada. So, it may be that bison know enough about wolves to appreciate their hunting patterns.

We've seen analogous behaviors in other ungulates. In Africa, vocalizations by lions and spotted hyenas are frequent. If zebras, wildebeest, or gazelles responded every time they heard a roar, their feeding might become so disrupted that maintaining healthy body condition might not be possible. They learn to ignore many of these sounds. Whether bison do similarly or rely more on their vision than on their auditory astuteness is unclear.

What to make of the differences between bison and moose? Other biologists noted that bison are a species of the open prairies and moose of the closed forest. It would be easy to claim differences in responses to predators arose merely because of habitat differences rather than differences between the species per se. In other words, if the situation were reversed and bison occurred in closed forests and moose in open areas, perhaps the results I reported would simply vanish; it is far easier to see in open habitats rather than when obscured by vegetation.

However, when I examined my data and accounted statistically for moose in open areas and bison in regions where visibility was reduced, species' differences still persisted. The results suggested that bison were less sensitive to the sounds of predators than moose. They said little about naiveté to predators, about the rate at which bison might learn, or behaviors that enabled them to survive the Pleistocene blitzkrieg while other species did not.

Clues about trust, fear, and culture would await pursuits in places far from Canada. For now, I'd move to species other than bison.

chapter 6

To Know Thy Enemy

> The hunters, in winter, find the deer congregated in "yards," where they can be surrounded and shot. . . . [I]n a panic, frightened animals will always flee to humankind from the danger of more savage foes.
>
> CHARLES DUDLEY WARNER (1878)

WARNER'S PIECE, FEATURED in the *Atlantic Monthly* more than a century ago, offers a seductively simple image of deer discriminating between types of nasty predators. Whether such recognition truly occurred is not entirely clear, but evidence 125 years later suggests deer and other species distinguish the severity of threat. This recognition dictates where species live.

As early as 1779, Canadian George Cartwright intimated that predators shaped the whereabouts of prey. Of the caribou of Labrador, Cartwright wrote they sought island refuges "to get clear of wolves which infest the continent." That was long before well-intentioned humans granted purposeful or de facto asylum for wildlife.

Today in Kenya, vervet monkeys tend to associate preferentially with humans when leopards are near, less so when they are not. Baboons share similar tendencies. But the tactics of using human presence to thwart possible predators is not unique to primates. Wildebeest in Zimbabwe's Hwange National Park rest on manicured lawns at night, a period when lions and spotted hyenas are most active yet still shy of people. Elk congregate around Yellowstone National Park headquarters, sites less frequented by wolves and bears. As in football, where running backs use lineman to buffer against enemy tacklers, prey have learned that associating with humans offers an effective defense. This behavior is hardly new. More than a hundred years ago, John Muir watched zebras and antelopes in Kenya, remarking, "Most of the animals seen today were on the Athi Plains and have learned that the nearer the railroad the safer they are from the attack of either men or lions."

The analogies do not cease with prey. Coyotes are no match for wolves, being regularly killed by their larger cousins. The innocent or careless may be the first victims. But in areas where they are neither shot nor trapped by humans, coyotes too alter their use of space, often selecting regions adjacent to roadways, areas wolves tend to avoid. Even cheetahs are averse to areas with hyenas and lions, but their vigilance does not cease with four-legged predators. When Maasai herders appear on the horizon, cheetahs flee. Maasai are well known for protecting their livestock from predators.

Rhinos and elephants in Kenya's Amboseli region have taken predator recognition one step further. During the 1970s and 80s they were speared by Maasai. The fierce warriors also placed loud bells on their widely ranging cattle. Researchers noted that both rhinos and elephants responded strongly to the clatter of bells but not to the cooing of cattle.

Back in the United States, as suggested by Charles Warner in his 1878 article in *Atlantic Monthly*, deer are also adroit when it comes to avoiding predators. During serious winters with high snowfall, white-tails from northern Michigan concentrate in forested regions known locally as "yards." Some are situated along the territorial borders of wolf packs. Border zones, whether defended by wolves or human armies, are extremely risky, because chances of serious aggression, injury, and death increase. By concentrating along such protected borders, deer reduce their wolf encounters.

This sort of behavioral avoidance is far from being restricted to deer and wolves. The Lewis and Clark expedition found wildlife more abundant along the transitional boundaries of different Indian groups. Known as "war zones," these were places where tribes, such as the Lakota and Mandan of the western Dakotas, were hostile to one another. The elk, bison, and pronghorn were docile and approached members of the Lewis and Clark team. They appeared less fearful than counterparts from other regions.

Descriptions like those of the 1805 expedition point not only to a lack of human persecution but also to curiosity. Inquisitive animals learn and may capitalize on that knowledge later in life. Those that are eaten as a consequence of bad decisions—like approaching a predator—will not get a second chance to pass along their lesson. If the cost of curiosity is less dire, offspring may also benefit.

Even today, heavily fortified military borders such as the 150-mile-long demilitarized zone that separates North and South Korea is a wildlife haven.

Despite the presence of a million troops, the 2½-mile-wide ribbon of habitat supports Asiatic black bears, white-naped and red-crowned cranes, and thousands of migratory waterfowl. Wildlife advocates have gone so far as to propose the creation of a transboundary international park.

MORE CONTENTIOUS THAN how the risk of death modifies a species' or population's behavior is its cultural emergence. How do prey acquire knowledge? How is fear transmitted? How much time is required for a population to adjust?

Answers have not come easily. Even with information, not all scientists agree. Sadly for wild species, including some of the world's most charismatic—pandas, gray whales, hippos, and giraffes—such information is generally unknown.

The word "culture" has more meanings than the world has continents. *Merriam-Webster's Collegiate Dictionary* defines it as "the integrated pattern of . . . knowledge, belief, and behavior that depends upon [man's] capacity for learning and transmitting knowledge to succeeding generations." For nonhumans, culture may be as simple as the acquisition of habits from others, which helps to explain why groups or populations of the same species differ. Given enough time and separation, it should be little wonder that populations develop their own trajectories. Although variation may arise, because some individuals copy the behavior of siblings, parents, or group members, others make discoveries by random exploration or through trial and error.

To understand how antipredator behaviors develop in a population is not easy and it is ever so challenging for large mammals. Researchers are faced with three initial difficulties—handling animals, obtaining adequate sample sizes, and performing field experiments. Although data may be amassed a point at a time, if samples are inadequate, there can be no statistical analyses. Without statistics, it is difficult to disentangle anecdote from fact, conjecture from pattern.

Large mammals pose another dilemma. Since body size and life span are related, species like elephants, rhinos, and chimpanzees have longer lives than porcupines, weasels, and mice. If learning is associated with age and a study's goal is to understand how information is transferred, a long-term commitment is needed both by individual researchers and funding agencies.

Among the ways by which prey recognize predators and pass such knowledge to other generations, several are obvious. There are observations and trial and error learning in addition to imitation. Genetics sometimes plays a strong role.

In fish, from African cichlids and tilapia to North American trout, components of predator detection differ. Some learn to associate odors with specific predators and modify their behavior accordingly. Conversely, predator recognition in Atlantic salmon is hardwired; the behavior is encapsulated in genes. For example, the newly hatched fry of wild Atlantic salmon that have no experience with their primary predator, pike, respond strongly to pike odors but not to other novel ones. Their reactions are part of a general fear response specific to pike that has little to do with learning predator recognition.

Young salmon are not the only species to maintain some entrenched capacity to fear predators. Captive mallard ducklings lacking in experience with the outside world show greater variation in their heart rates when exposed to silhouettes of hawks than those of geese.

What about humans and other primates—do they show a general fear response? The simple answer is yes, especially to snakes. Serpents cause fright. Toy snakes and toy alligators kindle more anxiety than toy rabbits or flowers, a reaction that suggests fear of certain classes of species runs deeply within our mammalian heritage. Wild rhesus macaques and squirrel monkeys also show stronger aversions to snakes than their captive-reared kin, but even those in confinement learn more quickly about reptilian perils than other sorts of dangers. Even though these responses are not governed by conscious control, individuals vary in their serpent-induced phobias.

For dolphins and orcas, elephants and chimpanzees, it has been difficult to detect whether a culture of fear exists. When information is passed among individuals via social learning, a case for cultural transmission can clearly be made—but only with the exclusion of other possibilities. A good measure, though not dealing with predator recognition, by which behavioral variation and the more general issue of culture may be gauged, is the notation of the occurrence of fads—the notion that some trends catch on for a while before fading into obscurity.

Over time, American fads have included wearing ostrich plumes, fur, and hooded garments. Others involve tattoos and earrings, dreadlocks and hair spiked purple or green, cowboy boots and goatees. Among nonhumans, fads are more difficult to detect. What, for instance, might we anticipate for a mouse or a zebra?

Animals fall prey to fads just like we do. When more than one animal in a group adopts the actions of another that it would otherwise ignore, a case for cultural transmission seems plausible. Sperm whales and orcas have recently been known to "steal" fish from long-line sets established to capture blue-nose grouper off the coast of New Zealand, tooth-fish from

Patagonia, and sablefish near Alaska. These actions occurred only within the same social units, although the events varied geographically. The behaviors had not been passed among groups, suggesting the development of specific traits within local groups, not throughout the population at large. Whether this behavior is just a fad or will persist is unknown, but it has adaptive value, since it presumably leads to more food.

I once detected what I thought was a fad among the wild horses in the Great Basin Desert years ago. A low-ranking mare from a specific band enhanced her social standing by lifting a sage bush in her mouth and shaking it at animals more dominant. In doing so, she displaced higher-ranking horses. Two others in her band developed the same behavior, something that never occurred among the other 135 animals from other bands in the population. The behavior occurred, however, in only one of five years and persisted for just two weeks. A different unique behavior has been reported in Indian tigers. A male learned to chase its prey, an elk-sized deer known as sambar, into water, where the kill rate was high. Opportunities for other tigers in that population to adopt similar behavior clearly existed, but none did. After the male died, the behavior was not reported again. Actions can transition into fads, but when the end point is beneficial, the appropriate moniker may simply be adaptive behavior.

The development of distinctive traits within a social unit or a discrete population with a clear link to culture is perhaps no better exemplified than in tool use. Orangutans, Asia's only great apes, show a fascinating level of creativity. One population on the island of Sumatra has figured out how to access the nutritious seeds of the Neesia tree, seeds which are protected by razor-sharp needles. The orangutans strip the bark and then use sticks to push out the seeds. Adjacent orangs have not figured this out even though Neesia trees also grow in their environs. Such knowledge is passed from one generation to another.

The most widely known evidence for cultural variation stems from our closest relatives, chimpanzees. At seven African sites where chimps have been observed for decades, in only one does combing of the hair occur. Sticks fashioned for fishing for termites occurs regularly in three. Overall, six of the seven sites have distinct patterns not found among chimps from the other areas, behaviors which typify unique developmental pathways.

In animals more distantly related than chimpanzees and other primates, the capacity to transmit information socially may flourish. Song sparrows, like other birds, develop local dialects that vary geographically. Killer whales sing differently from pod to pod. People from some parts of the planet wear turbans, others baseball hats.

More puzzling is how behaviors among solitary species spread through a population. Learning by trial and error or transfer of knowledge from mother to young are possibilities. Although my search to discover behavioral variation in bison was less than successful, I would pursue the possibility of learning in a different species threatened by predation. Its history has been almost as tragic as that of the bison, but it, too, has been rebounded.

APRIL'S SNOW WOULD turn slushy in the high mountains. It was still March along Colorado's Front Range. The ground was already bare, and the temperature in Denver would reach 65°F today. Pockets of green growth shimmered on south-facing slopes. I was a hundred miles from my destination, Rocky Mountain National Park. Its snow-capped peaks broke the horizon at fourteen thousand feet, but they were of less interest to me than the elk that roamed far below.

In recent decades, many had left the three hundred thousand-acre reserve for an easier life. Just as people flock to Aspen to enjoy the town's quality of life, elk moved to the town of Estes Park. With its plethora of plowed driveways, golf courses, and garden-laden trophy homes, this bustling tourist mecca had plenty of amenities for both humans and elk. Estes Park boasted the highest density of non-supplementary-fed elk in the world, up to a hundred animals per square mile.

The concentrations are likely anomalies of the past, particularly because some areas are now better havens than others. If elk remained in the park, movements through deep snow would drain precious calories, increasing the risk of death through malnourishment. Not only did the park contain deeper snows, but the protected reserve had been overgrazed during the last half century as the burgeoning population expanded.

In town, life was easier. Although elk contended with bike paths and skiers, Porsches and Humvees, the snow was less deep, people well behaved, and most dogs leashed. Spring's new grasses emerged earlier. Still, life was not perfect. Some elk died in collisions with speeding cars. Others became entwined among parked bicycles while looking for places to scratch their straggly winter coats.

Most important, elk in this part of Colorado's lofty mountains no longer worried about old adversaries. Grizzly bears were extinct by 1915, wolves 15 years earlier. Not only had these carnivores vanished, but so had the elk, all of them. They had been extirpated by market hunters eager for hides, meat, and antlers. The current population stems from a reintroduction of animals from Yellowstone, just after the last wolves were killed. If there was a place today where elk were tame, no longer buffet items for wolves and grizzly bears, and available for study, this mountainous part of Colorado was it.

To assess whether elk had lost their awareness of formidable predators, I would contrast their responses between areas and rely on the same methods as with bison and moose. Elk, however, offered a more prodigious challenge. They coexisted with other dangerous carnivores. Coyotes, cougars, and black bears ate not only elk calves but also the occasional adult. Whereas I previously confirmed that these smaller carnivores had little to no impact on the calves of moose or bison, elk were not immune. Predation by these predators has been substantiated during the last forty years.

I wanted to determine whether elk had lost, retained, or altered their reactions to the two large carnivores that had become extinct at the turn of last century. Sites beyond Yellowstone and Grand Teton parks, including all of Colorado, New Mexico, Utah, and South Dakota, had lost their grizzlies and wolves during the last 75 to 150 years. The problem was that most sites still retained coyotes, cougars, and black bears. If elk were invariant in their antipredator conduct, three possibilities existed. First, they might be similarly afraid of all predators smaller in size than grizzly bears and wolves. Perhaps they would just be reacting to predators in general rather than specifically to wolves and grizzlies. Second, sufficient time may not have passed for a relaxation in vigilance to develop. Finally, elk might lack the capacity to adjust their behavior to different predators.

To find out, I would again conduct playback experiments, focusing on females as I did on moose, since they are less sought after as trophies by humans. If nothing else, it would be interesting to see how elk behaved in areas of varied predation pressure and to understand what role, if any, wolves and grizzly bears played in maintaining behavior associated with memory and experience.

With my dog Sage and my field gear, a computer, and sound equipment packed into the VW, I rolled past Fort Collins. The suburbs vanished as quickly as agricultural pastures and sagebrush. Moving from the prairies to mountains, I sputtered to gain altitude. I looked at the passenger seat, half expecting to see Carol, as if we were traveling from Nevada to Wyoming. This time, I was by myself, a casualty of divorce after 18 wonderful years and a trove of memories. As I grappled to regain emotional control of my life, the uncertainty about Sonja's future loomed large.

The forests of ponderosa pines thickened; the sky was blue. In Estes

Park, I grabbed a motel room which did double-duty as a field office. On the lawn outside were tawny bodies with large white rumps.

THE HEADLINE—*Do Elk Fear Wolves?*—tantalized. It underscored a more fundamental issue, the meaning of fear itself. While both animals and humans display some characteristics associated with fear, such as vigilance, trepidation, and anxiety, the more familiar framework used by psychologists and neurobiologists is the "generalized fear response." It refers to a specific set of events that begins when a sensory signal causes fright and stimulates a part of the forebrain called the amygdala. This, in turn, triggers the hypothalamus to produce a corticotrophin-releasing hormone. The adrenal gland subsequently kicks in, secreting cortisol. The brain and muscles are readied for danger. Remarkably, since the forebrain is signaled even before the cortex is stimulated, fear can be aroused even when there is no danger present.

When an elk, or another animal, uses its mélange of sensory equipment—olfactory, auditory, tactile, or visual—to assess its immediate environment, it is often inventorying more than danger. Food, mates, kin, annoying insects, and shelter are also important elements in one's immediate environment. Just as humans may be wary for many reasons, so too may animals.

Predation is just one of many possible activators of fear, but it differs in a critical way. Feeding on the wrong bush or walking an incorrect trail will not end one's life. Predation will. Vigilance can pay large dividends until individuals become so hypercautious that they no longer feed efficiently. Balance is the key. Gazelles in the Serengeti pay a high price for wariness—they lose calories. The cost is steeper for those who are not wary; cheetahs eat them.

At Rocky Mountain National Park, the senior government biologist reviewed my experimental protocols. So did the park police, who were informed that I would amplify fake sounds through a speaker to discover whether the park's naive elk had any recollection of wolves or grizzly bears. Initially skeptical, they agreed that if wolves did arrive in Colorado, it would be advantageous to predict elk behavior. I promised to work early, rising well before most tourists. My permits were approved.

During the next five days, I hiked, snow-shoed, walked, and climbed in search of ideal playback locations. Most were for audio experiments, but I also planned to use odors, since these elk were not afraid of people. As I worked, Sage sat patiently in the van. She had grown accustomed to the smells of bear scat and wolf urine that perforated the containers.

She no longer responded to coyote yips or wolf howls. Elk, however, were a different matter. When detected, Sage would thrust herself against the windows, hoping to escape and enjoy a primeval pastime: the chase. Fortunately, the elk rarely saw her and she never left the car while within the park's boundaries.

The first "wild" elk I encountered numbered eight in total. They fed on desiccated grasses along a stream where I imagined beavers were once abundant. The snow cover was spotty, four to six inches deep. On a knoll about half a mile distant, I set up my equipment. First, I played the sounds of running water, then howler monkeys, then coyotes. An elk at the far edge of the group was the first to look up, followed by her nearest neighbor. They remained vigilant for about thirty seconds. The others ignored the sounds. I continued the inventory—ravens, red-tailed hawks, hyenas, tigers, and wolves. The elk were unimpressed. About an hour later, people had discovered my van and realized that the tiny dots in the distance were elk. These elk were too far for a comfortable walk. The tourists left.

Down the road, I discovered a different valley, pulled over, and began hiking. This time I found a larger group, 35 animals only 400 yards away. I assumed they were habituated to people but needed to use the same methods as before. I skulked around silently until reaching a large granite boulder that concealed all but my speaker. Again, I repeated the playbacks, this time varying the order of sounds. The results were no different. Most animals fed nearly 95 percent of the time. As before, the peripheral animals were more apt to stop feeding, but these tangential foragers fed 80 percent of the time.

At 10 AM, I left the park and pursued elk in town and along the suburban river channels and forest edges. Their responses mirrored those of the park's animals. I also tossed snowballs tainted with predator scents. While the odors triggered some curiosity, none of the animals became alarmed or ran.

By midday I was done, as elk enjoyed their afternoon lull. Some lay ruminating in the open. Others napped on the golf course. I returned to the motel and entered data before taking Sage on a jog. Each day I repeated the routine. By the time I amassed 160 data points, it was clear that the elk in this part of Colorado were not concerned with predators. Even if coyotes or cougars were eating elk, wariness was not prominent among their behaviors.

Perhaps the elk in this part of Colorado were unique. This was but one study area. If elk learn, then it was no wonder that the animals here shared a disinterest in predators or novel cues. Since immediate danger was slight, why bother wasting time on surveillance when food at this stressful period of the year was tantamount to life? To avoid starving, elk might need to feed continuously.

As attractive as this explanation was to account for their high rates of feeding, it was also the reason why my experiments were performed for all species at the same time of year—late winter and early spring. In seasonally cold regions, all animals confront similar environmental challenges. The difficulty of finding food at winter's end means that individuals use their body reserves to sustain themselves, especially because vegetation at this time of the year is low in protein and offers little nutritional value. Elk, along with bison and moose, are programmed to subsist on their body fat. Just as most bears hibernate, northern ungulates have evolved a similar tactic. They too lower their metabolism to survive the harsh months. There is one large difference. Whereas elk, bison, and moose just slow down for winter, bears restrict all actions to a single spot: their dens.

Beyond food, there is another possibility for why elk in this part of Colorado were nonresponsive to potential predators. In Estes Park, animals were bombarded by cues associated with humans. They heard dogs and smelled diesel-spewing trucks. They saw people walking, running, and biking. They heard airplanes, helicopters, and ambulances. These elk lived in part of an American society filled with action. If elk had habituated to such an environment, they might fail to respond, not because predators were unimportant, but because people were overwhelming.

Habituation is the waning of a response due to repeated contact; it does not result from age, nor is it followed by any type of reinforcement. If an animal is afraid of humans and runs but is never chased or threatened, over time it may become more tolerant of human presence. The familiar adage that familiarity breeds neglect but rarity engenders reaction helps us understand something about the peril and advantages of prey habituation to potential predators. If prey do not become attuned to the potential lethality of a rare event, then an infrequent but successful attack will remove that individual from the population, and its genes will not be replicated in the next generation. The unobservant will be removed with the resulting survivors being slightly more wary.

Besides elk, I also considered the possibility that moose in the Tetons might be more habituated than those in Alaska. If a failure to respond to predators was a corollary of habituation to people, then moose should differ in responses based on their association with humans and not with predators. However, this was not the case. Moose from the remote Talkeetnas in central Alaska, where they experienced few people, were just as alarmed as their cousins from Denali National Park when encountering the cues of bears and wolves. Similarly, the degree to which bison were habituated had no effect on their lack of responses to all predators except African lions.

The use of different study areas for elk would also allow me to adopt a comparative approach to understand habituation. From Colorado I went north to areas of Wyoming where elk were used to people and to areas where they were hunted by people. Using such sites would allow the contrasts I needed to pursue the question: do elk fear wolves?

THE IDEA THAT prey are alarmed by carnivores and that their ecological communities are shaped by these reactions is not new. Georg Stellar, the talented naturalist of Vitus Bering's 1740s North Pacific expeditions, noted how humans played the role of apex predator. Stellar explained why sea otters were found on some shorelines but not others. Those found along Alaska's Aleutian chain were unafraid of people, because they had not, at that point in time, been hunted for their fur. This was not the case along the coastline of the Kamchatka Peninsula, where they were pursued by Russian furriers. The wary otters were further out at sea.

The converse can also be true. Although we humans are the ultimate predator today, we too have assuredly been prey and remain so in a few areas.

Our past fears of large carnivores may have had important ecological consequences. Some fifteen thousand years ago, short-faced bears must have governed the movements of Paleolithic peoples lacking sophisticated weaponry. It is possible that these now-extinct bears, twice the size of modern grizzlies, hindered our ancestors from crossing the grassland steppes that once connected the shores of Siberia to Alaska's Seward Peninsula. If so, then the threat of predation delayed colonizers of Asia from entering North America. While this is admittedly speculative, if true, then our fears of being eaten may have stalled for several thousand years the ecological destruction we subsequently wrought as prima facie hunters upon the naive North America fauna that evolved without us. Whatever truly occurred twelve to fifteen thousand years ago has relevance for understanding some of the resulting fear that stems from our conflicts with carnivores in our contemporary world.

Few Americans fear predators simply because most of us no longer live in areas where we can be part of the food pyramid. Beyond our borders, peasants living in the midst of big, dangerous animals may still pay the price of their impoverished lifestyles. East Africa is one such place. Two male lions struck terror in Kenya where, in 1898, construction of a railroad line between Mombassa and Lake Victoria was well underway. More than 130 workers were killed, as first described by John Patterson in *The Man-Eaters of Tsavo* and later brought to life in the 1996 film *The Ghost and the Darkness.*

In parts of southern Tanzania, a legacy of fear continues. Between 1932 and 1947, a total of 1,500 villagers were reportedly killed by lions. More than fifty years later, lions also turned intimidation to death in the hamlets of Simana and Nusuru. During 2001 and 2002, there were 22 attacks that resulted in 14 human fatalities. Understandably, people worry. In 2005, one villager reported, "This is what happened the last time. The lion ate the dogs and goats first, then he began eating people."

Out west in America, worries still concern the presence of big predators, but not because of the peril of predation. Some sportsmen are concerned that the prey that might otherwise be put on the table as food will be consumed by cougars, wolves, and bears. Since hunting by humans remains a pervasive global activity—in the United States, about one of every seven males between the ages of 16 and 65 has hunted—it makes sense to understand our past as hunters.

Over the course of history, some fifty billion people have inhabited the Earth. Ten thousand years ago, most societies harvested meat. Five hundred years ago, only one percent did. A propensity to eat wild game has not only changed inversely with human population size but, predictably, so has the abundance of native large carnivores. For some the loss is a justifiable progress, because a world in which predators are scarce is also a safer world.

Might hunters claim that the removal of predators as competitors inevitably increases their success at game harvesting? Surprisingly, the answer may be "no." If a decrease in native predators enables prey populations to achieve higher densities, an increased availability of additional hunting permits may offset hunting success. With high levels of human hunters, the per capita success of each may be diminished, because there is only so much wild game to go around. Instead of viewing predators as our rivals for prey, an inevitable result of their loss might simply be increased competition among humans.

Moving further into our distant past, there must have been great antagonism between humans as hunting primates and carnivores. Although the role of meat in the diet of monkeys and its relationship to human origins is controversial, African primates, such as bonobos, chimpanzees, and baboons, will kill and eat juvenile ungulates. In doing so, competition with some African carnivores certainly seems possible.

More salient is the extent to which our more recent post-Pleistocene ancestors selected the same prey as contemporary carnivores. Cave paintings and drawings etched in rocks by humans ten to forty thousand years ago provide evidence of great respect for mammoths, bison, horses, rhinos, ibex, pronghorn, caribou, and other species. Undoubtedly humans consumed many if not all of these species, as did other predators.

Competition for food between humans and carnivores is also known even today, and it heightens the reality of fear and risk of death. In Uganda, starving peasants are regularly displaced or even killed when attempting to expropriate meat from lions and leopards. In the Russian Far East, the converse has been true. Tigers have been displaced from their kills by hungry but better-armed humans who subsequently consume the meat. The certain level of dietary similarities among contemporary humans, ancestral humans, and carnivores when food may have been limited suggests that competition between "us" and "them" has been part of our *Homo sapiens* heritage for a long time.

ALTHOUGH THE TETONS were shrouded, it was not the customary clouds that obscured the peaks. There were dwellings, trees, people, and power lines. I had moved to the tiny settlement of Kelly. Bounded by the National Elk Refuge to the south, Grand Teton Park to the north and west, and national forest land to the east, this enclave of two hundred souls had to be amongst the most eclectic in Wyoming. Within the community was a cluster of yurts inhabited by free spirits from the sixties and newcomers looking for adventure. There was a physician from Nepal who once treated a friend who had been gored by an Indian rhino. There were climbers who scaled peaks in the Andes, and others who had done the Himalayas. There were bikers with BMWs and bikes for mountain trails. There were writers and rangers, a veterinarian and a Nobel laureate, llamas and a goat rancher. A mere eighth of an acre of land listed for more than $330,000. The nearest stores were a dozen miles away.

From my window, I watched the elk that dotted the wintry landscape, eleven cows in one group, six antler-adorned males in another. A moose rested on the adjacent hill, and two moose cows with last year's calves were only a hundred yards distant. A lone bison bull fed above the warm springs. Two bald eagles roosted in cottonwoods. The thermometer read –12°F. While not quite Africa, Kelly was not bad.

Small-town Wyoming, even Kelly, was not for everyone. A New York writer called one day to arrange a face-to-face interview. I explained directions. "Turn six miles down the road from the junction. It will be snow packed. Once you hit the cattle guard, turn right; my office will be third on your left."

"This sounds exciting. I'm confused by one thing. I know what crossing guards are for school kids. In Wyoming you must do things differently because of all the cattle. I don't plan to hit the guard. If I come in late, will there be no cattle guard?"

Perplexed by her response, I paused. *What was so confusing?* It dawned on me. The writer expected the guardian of the cattle to be a person, not a metallic grating on the roadway.

A MEETING WAS scheduled to discuss how the federal government and researchers would evaluate wolf effects upon wildlife. The US Fish and Wildlife Service (USFWS), the sole management authority over the reintroduced wolves, wanted yet another formal proposal, the third in two years. In attendance were Bob Schiller, the chief of science for Grand Teton Park, Steve Cain, the park's senior biologist, and Mason Reid, the coordinator of the park's wolf program. Also present was Sanjay Pyare, a postdoctoral scientist from the New York–based Wildlife Conservation Society. Mike Jimenez, Wyoming's field representative for wolf recovery, represented the USFWS.

I had just secured a $198,000 grant on behalf of the park, with the goal of evaluating ecological effects of wolves on prey species in Jackson Hole. This meeting was to iron out specific responsibilities, something that ordinarily would have been done well before now. But studying wolves and elk in the northern Rockies was anything but ordinary.

Mike already knew this well. He had just spent forty minutes explaining the diatribes of anti-wolf advocates, his face-to-face meetings with vein-popping, red-faced, angry men only inches from his nose, and on-the-ground encounters involving pistols and other hostilities. There was the time at the high school gymnasium in Salmon, Idaho, when a sheriff tackled someone from the audience who attacked the stage as a USFWS biologist calmly explained why the American people overwhelmingly supported wolf reintroduction. Despite my grant, Mike was disinterested in any form of collaboration. The complexities of working under multiple jurisdictions in Wyoming could mean only one thing for him—more headaches.

"Politics override everything, which is unfortunate, but I'm not interested in having data at that cost."

Others in the room tried to convince Mike otherwise, that the public benefits by understanding wolf-prey relationships, that the study would cost the USFWS nothing, and that some data are better than none. I tried to prevail, asking where common ground existed and whether he might be interested in collaboration if we offered aerial support, GPS collars, or money for wolf captures. "What about a person on the ground with your group, or perhaps to find out kill rates?"

Mike answered "yes" to each. But his position quickly shifted into a series of negatives. We all haggled for an additional thirty minutes. Finally, he slammed a fist on the table, exclaiming our need to understand that the USFWS was not interested in collaboration or interagency cooperation. There was a single goal—wolf recovery—and our interference was not going to help. Mike was legally in charge, and the final word was his.

I was disappointed. So were the park's staff. These were federal research funds that I had written proposals for on behalf of the park. The money was awarded from regional government offices in Denver, with final approval from Washington. It was ludicrous that a different federal group, one that had supported research on wolves elsewhere, would totally oppose academic involvement or that of another federal agency.

At an individual level, Mike's position was understandable. He was not only a professional biologist but also a person with feelings. Yet none of that was clear in a local news story, where the headline read "Government Agent Kills Wolf." In the same story, Mike's home phone number was posted. He was damned. Environmental groups wanted a piece of him for killing wolves. So did the state and many local ranchers for not killing them.

Four years later Mike, was arrested and charged by the State of Wyoming for trespassing and littering. The incident involved the capture and collaring of wolves from a helicopter. For safety, the immobilized animals were removed a few yards away from the road where they had been darted and placed on adjacent private property while recovering from anesthesia. The wolves themselves were the "litter." A federal judge dismissed the case as groundless.

The politics of wolves are daunting, replete with irrational games and power struggles. I was not so naive as to think that studies would be easy, but I was more than sickened by the flexing of muscle to blatantly suppress the acquisition of knowledge about wolves.

The public understandably had an expectation that the government would work on their behalf. Federal biologists were not excluded from that belief. With hard won, well-intended federal monies, my Park Service colleagues and I were unable to sway the Fish and Wildlife Service to cooperate. Mike wanted no part of a project to understand wolves in Jackson Hole. Did he or his bosses seriously believe that the American public would be better served by an absence of information?

WEEKS PASSED. ELK became increasingly visible on the ten thousand-hectare National Elk Refuge. Created in 1912, its purpose was to sustain the elk of Jackson Hole, a region where many migrations had been severed and habitats lost. At times, more than eleven thousand elk crowded into the fields that border the town of Jackson. Once wolves arrived, however, they garnered the attention.

Not only were wolves observed chasing and killing elk, but tourists and locals alike watched copulating pairs on different days in February. Then, a cougar and her three kits arrived. They too dined, usually on deer, but sometimes on elk. For six weeks they fed in the open as people viewed with delight from the roadside. Other than Jackson Hole, there was no place in the world where one could sip a cappuccino and watch large predators eat elk in the morning, ski in the afternoon, and listen to wolves howl at night.

While wolves made their presence known in the southern part of the ecosystem, I drove 350 miles north to Lamar Valley. Situated at the edge of Yellowstone National Park, along Montana's boundary, elk had already been exposed to wolves for more than three years. Doug Smith, the park's gifted carnivore biologist who found time to cooperate with outside researchers, found that elk comprised nearly 95 percent of the wolves' diet. If elk were going to stave off localized population crashes, they needed to learn quickly about their powerful reintroduced nemesis.

Yellowstone National Park was not only the ideal place to contrast elk knowledge of predators with those from Rocky Mountain National Park; the direct contrast would help support or refute another component of predator recognition. Since elk from Yellowstone were the single source for restocking at Rocky Mountain Park, the same genetic lineages were at each site, one with wolves and one without. If, like Atlantic salmon, an innate component to predator recognition existed, then elk from both locations should share a similar response to predators. Either elk from both parks would respond or they would not. There was no middle ground. If elk from one park reacted one way, but those from the other park behaved differently, then it was likely that this variation in behavior was not genetically hardwired.

There was yet one additional reason why Lamar Valley was so attractive. It had little to do with its vastness, its native grasslands, or its spectacular diversity that mixed spotted frogs with curlews, boreal toads with flycatchers, and craggy peaks with spruce forests. It was here that I watched grizzly bears search for, chase, and catch newborn elk calves. It was here that other biologists documented grizzly bears killing adult elk and consuming perhaps forty percent of the calves born in a single year. There was just no place better than Yellowstone to expect bears and wolves to affect elk behavior.

In the park, I worked from early in the morning until it was too dark to see. Outside the park, I was more opportunistic, conducting playbacks wherever and whenever I found animals. On the outskirts of Gardiner, Montana, animals were habituated and so I presented them with the smelly morsels of grizzly bear feces, wolf urine, and other odoriferous gems.

In contrast to Colorado's elk, those from northern Yellowstone were high-strung. When alerted by the caws of ravens, they become vigilant, sometimes squeezing together in tight groups for safety. Before clustering, they moistened their noses and tilted their nostrils upward in an attempt to deploy all senses while surveying for danger. The only difference in response by Yellowstone elk to raven caws and wolf howls was that when the animals were in thick vegetation, they fled more frequently from the howls.

Because the variation in elk behavior toward predators was slight, I decided not to stress them further with additional sounds. After three days, it was clear that these elk had plenty of angst. While the scientist in me yearned for a better sample, my heart rested with the poor animals. Yellowstone elk now lived in a landscape of fear. Fleeing from me was not an additional burden they needed. My work ended.

In the process, I had also learned that the context of olfactory playbacks mattered. The elk were far more responsive to the sound cues of predators than they were to odors. Unlike moose, which in Alaska were sensitive to wolf urine and bear scats, I was unable to produce strong countermeasures to predator odors in elk. Whether the disinterest reflected biological reality or was an artifact of my experimental design was uncertain. With moose, I conducted the scent experiments both in wooded and in open areas. For elk, playbacks were in open areas only, places where I could approach, but these were also the areas where elk could rely on their vision. Perhaps, in more closed environments, or at other seasons, females might have been more responsive.

I now knew that elk behavior varied in response to wolves. I did not know how antipredator actions developed or how they were modified. I did not know whether hunting by humans played any role at all or if it exacerbated fear of nonhuman predators. Social learning and trial and error were both possibilities for the differences in behavior between elk from the mountains of Wyoming and Colorado.

Without radio collars on elk, I would know nothing of the history of individuals—their ages and personalities, their geographic overlap with predators, and whether their offspring had lived or died. My thoughts about possible causation would be uninformed, mere speculation. The bottom line was that I had no idea what individual elk knew or learned. I knew only about elk from different regions but not if or how learning might progress among individuals. My attention refocused on the moose I knew best, those that wore collars and had names and personalities—those I had followed in and beyond the Tetons.

chapter 7

Among the Naive

> A mother's job is not to provide an impervious protective bubble around her child, bur rather to teach her child safe ways of dealing with the real world.
>
> MARTY CRUMP (2000)

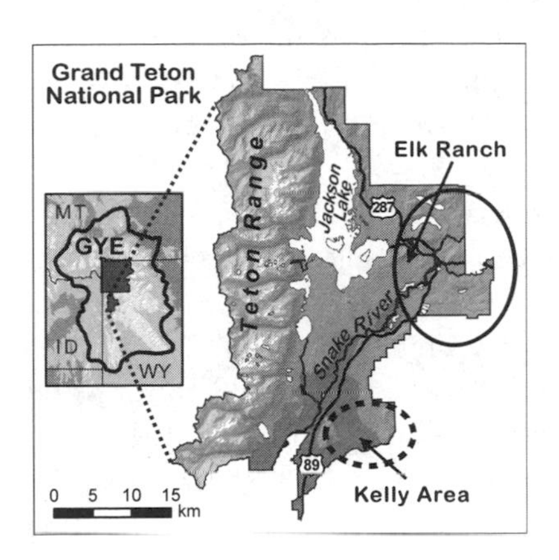

CNN CARRIED THE headline "Yellowstone Moose Decline due to Hunger, Not Predators." Despite the title, the 1999 story focused on animals in Grand Teton. Mothers and calves were thriving, but unexpectedly, the number of calves was relatively low. Pregnancy rates in adult females had declined over the past thirty years. Currently hovering about 75 percent, pregnancy had averaged more than 90 percent when Doug Houston completed his doctoral research on the same population in the 1960s. The recent decline dropped the Jackson Hole moose population to the bottom tenth to fifteenth percentile in North America. A second marker of fecundity was also low. Like deer, moose have the capacity to produce twins, an event which occurs only when mothers are healthy with much body fat. Alarmingly, twinning rates in Teton moose had also diminished.

I wrote an article detailing these statistics, which was published in the journal *Conservation Biology*. My coauthors and I speculated that moose may have reached the point where food was limiting reproduction. If females were in poor condition, their chances of becoming pregnant were lower. Even if we were wrong about the cause of low pregnancy, wolves and bears were not killing the majority of neonates. Calf survival remained high until animals migrated back to their winter ranges.

No one doubted that wolves were killing elk, but evidence of predation on moose was lacking. The ominous forecasts of a moose slaughter had not materialized. This was understandable, at least to me. With thousands of elk in Jackson Hole, it made little sense to expect that wolves would risk injury to take on a more dangerous quarry.

In other areas, including sites in Alaska and Michigan, wolves have had appreciable effects on moose demography. Where moose were the only or the major prey, wolves had little choice and took their chances while suffering broken ribs and legs, cranial injuries and death. Given the abundance of less dangerous food-on-the-hoof, wolves might be prudent not to attack moose.

In Jackson Hole, elk number more than ten thousand. Most are supplemented during winter with rations of alfalfa pellets. Concentrations at the six feed grounds reach as many as 1,200 elk per square mile. As a consequence, adjacent plant communities pay a high price. Stands of serviceberry and chokecherry are pummeled. Aspen, willow, and cottonwood trees are heavily browsed. Local forests have disappeared, along with the birds that depend on them. The overall elk population is so large that the Wyoming Game and Fish Department generally issues far more hunting permits than are filled during most years.

While hunters are disdained by many in the animal rights community and often misunderstood by those from urban centers, hunting is deeply rooted in our culture, and most modern hunters maintain profound respect for animals. Michael Soulé, the first president of the Society of Conservation Biology and known popularly as the father of that scientific discipline, offered a poignant perspective on the value of hunting.

> One of the things I like best about hunting is that it's humbling. There I was professionally successful, my work widely recognized, which left me feeling pretty good about myself. But out there in the woods, it was a whole new game. I was a child again, a beginner making all kinds of mistakes, lots of errors in the presence of game: moving at the wrong time, making too much noise, "buck fever." Being humbled by nature, as opposed to being humiliated by people, is a deeply valuable experience. All you can do is laugh and try to do better next time. And as a biologist, hunting provides me with an opportunity to experience what Zen practitioners call "beginner's mind": to learn about animals directly from animals.

ONE DAY THE wolves arrived. Within a few weeks, they had discovered not only the National Elk Refuge, but other feed grounds to the east, up the Gros Ventre River beyond the tiny village of Kelly. Often before first light,

I'd work the roads looking for deep-toe imprints. After a fresh snow, the tracks of an entire pack were highly visible. By following them, it soon became obvious when moose were encountered. Instead of tracks with big paw prints, vertical holes deep in snow dropped two to three feet. These could have been made hours before or after wolves were in the area, and they did not say much about an interaction. Other evidence did. If gaits changed, snow would be more scattered and the distance between steps, both of wolves and of moose, would increase. If wolves were in pursuit, tracks would go in the same direction, and the wolf prints veered from roadways.

Because wolves were active during morning at this time of the year, it took little time for my human neighbors to detect them. Two traveled together within a hundred yards of Kelly. Others were in a nearby pasture with Texas longhorns. The radio-collared moose used adjacent gallery forests along rivers still thick with willows and cottonwoods.

Although I hoped to witness interactions between moose and wolves, I could not afford to sit, wait, and watch ten hours a day. Other chores needed attention. On some days, I directed playbacks to known females so that I could understand their responses to wolves, ravens, grizzly bears, and tigers, as well as to neutral scents or sounds. Just as parents know the behavior of their children, my intent was to understand each moose. That way, I'd learn about differences among individuals and the degree to which their behavior changed over time. Other days, I collected fecal samples to determine who was pregnant and checked the status of all animals. If calves had died, I wanted to evaluate how they differed from those who lived. Might survivors be larger?

My interests were more than academic. The Game and Fish Department permitted the harvest of female moose in autumn, but information on the subsequent survival of calves without mothers was unavailable. If orphans had poor prospects, then the single bullet of a hunter seeking a cow would kill two animals, both mother and child. If I could gather data on calf survival with and without mothers, I could offer information about juvenile survival that would inform hunting policies.

To find out whether size mattered, I needed to estimate calf dimensions. Rather than handling animals—which involved risk and expense and was not especially a good experience for either the animals or the park visitors—I took photos. By measuring the distance between calf and camera, I could modify a technique I once used to evaluate horn size in wild black rhinos. In this case, my interest centered on calf head size, since there was a strong relationship between head dimensions and body weight.

My goal for the Tetons was slightly broader. I was going to photograph every calf of every known moose mother to estimate their weights in relation to survival over the winter. Shots were needed from two angles. The first would be a calf facing the camera directly so I could measure the linear size of the face, from eye orbit to orbit. I also needed a profile to determine the distance between its nose and the distal point of its skull, the occipital condyle. By combining the two measures, I'd have an estimate of head size that could then predict body weight. The technique had appeal, as it required no animal handling, but it still carried liabilities. With a 300-millimeter lens, I needed to be within thirty yards for accurate measures (digital photography was still a few years away).

During the first four winters, my crew and I accrued information on about seventy calves. Of those that perished, virtually none had fat on their ribs or in their bone marrow. Whatever ancillary energy they once had was gone, slowly catabolized. The calves wasted to death.

One particular female calf offered encouragement, not only because of her drive to survive but because she revealed insights about her relationship with her mother. This calf we called Zena. She was only eight months old when her mother, Zen, was killed by a speeding trucker. Zena was at Zen's side when she died, and the juvenile faced an uncertain future.

The process of becoming an orphan cannot be pleasant, irrespective of whether a youngster is part of a social group or just associates with a mother. We know from studies of rhesus monkeys conducted fifty years ago that young deprived of contact with their mothers attach to bare wires or rags left as surrogates in their cages. These orphans failed to develop "normal" behavior. They were more fearful, less socialized, and less likely to be good mothers themselves. The contact with their mothers was important for normal development. Although no one has attempted to understand long-term effects of orphaning in moose, Zen's unfortunate death offered an opportunity to get to know her little orphan better.

Zena remained at her mother's carcass for two days, some two hundred yards from the highway. It was February, 1998. Zena watched as her mother's frozen carcass dwindled under the voracious appetites of two dozen ravens, a bald eagle, and a pack of coyotes. Without her mother, Zena's only source of protection and knowledge vanished.

Two weeks later, all that remained was a desiccated hide, a skull, and some reddened bones, buried under a fresh coat of snow. That's when Tom Roffe and I decided to radio-collar the recent orphan. Sinking to our waists in snow, we approached Zena on a –14°F morning. Terrorized, she ran, plowing through chest-deep snow. Then, she stopped suddenly, as if the spot had some magical quality.

As Tom fired a dart, something caught my attention. We had lost our orientation. Zena stood above patches of hair, blood smears, and skeletal remains. She had run more than half a mile from us only to stop at the precise location where her mother died. Perhaps she sought protection, hoping that somehow her mother would rise up and drive away the dastardly researchers.

During the next three months, we followed Zena, tracking the signals emitted from her radio collar. Defying our expectations, she was still alive in April. By May, spring was palpable. The sun shone and the air carried warmth. Watery rivulets were everywhere. Sandhill cranes had arrived, and the melodies of song sparrows and robins broke the early morning silence. Willows had buds.

Unlike the other adults and calves that had already migrated to summer grounds, Zena remained behind. Without her mother, she did not know where to go. But she was alive, and my crew and I celebrated, since the rich emergent springtime growth would offer protein. Two weeks later, I detected a rapid pulse on my radio receiver. It was a mortality signal. Zena was dead; the cause, malnourishment.

In addition to starvation, vehicles, and hunting, moose died for other reasons. One collared female perished in a spring avalanche, but it took another month to find the body and confirm the cause. And, while no death can be pleasant, some are more brutal than others.

In January one year, when I tuned my receiver to 151.673 megahertz, the signal blipped rapidly. A nice friendly cow called Juniper had died. Her unfortunate death created the opportunity to determine how her seven-month-old calf, June, would fare. The signal came from a patch of leafless willows surrounded by ridges and hills of rolling sage. The snow was about two feet deep. After three hours of tromping, I narrowed the search to a pool of frozen muck about 10 feet wide and 15 feet across. A large open hole was in the middle. From the edge I probed the jumble of mush with my ski pole. The ice did not hold, and as I plummeted into the quagmire, I hit a stout leg. Juniper was entombed below. She too had fallen into the natural sump, and must have succumbed slowly. By March, her calf June had also died.

I tallied my results and reported them to my friends at Wyoming Game and Fish. Their interest piqued. They had no idea that orphan survival was so low. Only one animal out of sixteen—a mere six percent—had survived orphanage. This contrasted sharply with the then current sixty percent survival rate of calves with mothers. Also surprising was that calf size had no effect on over-winter survival. All that mattered was the presence of one's mother. Without mothers, calves did not know where to migrate, how to locate the best food, or how to avoid confrontations with other moose.

Calves, even those weighing up to 350 pounds, were consistently pummeled by more dominant animals. After a bevy of unsuccessful attempts to feed, they either left on their own volition or were driven away. Some calves wandered alone until death, others tried affixing themselves to different mothers. In the end, with eyes vacuous and bodies weakened, the solitary calves could be approached quite closely. Without strength to run, they would soon move no more.

Not all calves were bereft of power, and not all died because they were motherless. In 1999, the first full year after wolves recolonized the valley, the calves of radio-collared mothers began to vanish. Their disappearance could have been the result of many factors, but some saw only one explanation: wolves. I considered an alternative: malnourishment. I would be dead wrong.

Like fisherman anxious for their daily ocean catch, the local wolves were hungry commuters moving miles back and forth between the elk feed grounds. Along the way, they encountered moose.

The first sign of death came in January when I noticed a dozen ravens picking apart something on the ground. Large canid-like tracks converged on an area with the carcass of a big moose calf. From its head size, I estimated the weight at 330 pounds. Nearby, two beds were melted in the snow, one large, the other small, a resting place of the radio-collared mother and her calf. From the smaller bed a blood trail led to the fresh carcass. There had been no chase, just a swift approach, and a take down. If the mother had defended the calf, the signs were absent. There was no snow scattered about. The mother's strides were in a different direction. It seemed likely that neither the calf nor cow knew that these large dogs were not coyotes. Neither they nor their recent ancestors had seen, heard, or smelled real wolves, for the predators had been gone from this valley since the 1930s.

Days later, six wolves worked the snowy sagebrush flats beyond Kelly. They had been there during the night and again early the following morning. I suspected that if the wolves had remained for 12 hours, there must have been a kill. With miles of tall sage and bitterbrush to search, the chances of me finding it were remote. If I did no searching, however, the outcome would be a given—no carcass. If I persisted, I might still not know, but at least there would be a possibility. So, on a warm 20°F day, I grabbed a spotting scope, climbed a hill, and prepared to sit until dark. My immediate goal was to look, stare, and search every inch of ground for something that might have caused the wolves to remain throughout the night.

By 2 PM, I was cold and giddy. A wind blew from the southwest. My eyes hurt. About a mile away I noticed a raven flying. Before pinpointing its location, it disappeared into the vast ocean of sage. Twenty minutes later another flew. This time I was ready and mentally defined the spot. I awaited more birds but none came. By 3 PM, I was seriously cold. With little to lose and anxious to warm up, I hiked to the spot.

As I neared it, wolf tracks appeared. Eight ravens then flushed from a bloodied carcass. I inverted the skin, and saw fresh punctures and hemorrhaging. The sequence of events was remarkably similar to the prior kill. A calf had bedded near its mother. Wolves approached to within a few yards. I surmised that the calf had not fled, at least far, and, judging by the intact ground, there had not been much of a struggle. The naive calf had absolutely no chance if its mother had mistaken these dogs for coyotes.

Throughout winter, I amassed evidence of eight calves killed by wolves, six from radio-collared mothers. Two points were especially interesting. First, the kills occurred in the southeastern edge of the park around Kelly, the area between the federal elk feed grounds to the south and the state feed grounds to the east. Calves were killed on wolf travel routes, not in areas where wolves generally hunted. Second, the radio-collared mothers that lived some ten to twenty miles north—well beyond the winter home ranges of wolves—enjoyed calf survival rate of ninety percent. In essence, moose calves were surviving outside of, but not within, the wolf's range.

The regional variation in calf deaths offered an opportunity to understand how or if moose develop responses once their calves are killed. Results from the playback experiments, and now direct evidence of wolf predation, made it clear that calves and mothers were ignorant of predators. Alaskan moose, by contrast, had been highly responsive to carnivore cues—even ravens—in areas with wolves and grizzly bears. The natural experiment I had waited years for had just occurred in Grand Teton. Now, I could assess whether mothers that lost calves would develop traits similar to their Alaskan cousins. If moose were to learn socially and possessed any sort of rudimentary culture, the opportunity to find out was here and now.

Elsewhere, biologists had characterized relationships between mothers and young but not when carnivores had recolonized a system. Some insights had nevertheless been gained about variation in the behavior of moose mothers. In Alaska, some remained with or protected their dead calves from predators for up to three days. Other mothers offered little resistance after a calf had been captured. Adolph Murie recounted maternal concern but an unwillingness to attack:

> On May 24, 1939, I approached a cow with a calf that was less than a day old. When the cow ran off, the calf followed shortly. Part of the time the mother trotted slowly enough for the calf to keep up, even though I was hurrying after them about seventy-five yards behind. Each time the mother found herself a short distance ahead, as happened four or five times, she returned and nuzzled the calf. When it finally lay down, the cow stopped about fifty yards away and did not run off until I captured the calf.

Although interactions between mothers with very young calves and predators have not been described from Wyoming, mothers demonstrate alarm and distress. I once observed a female whose calf was hit by a car. She refused to leave the site until a front-end loader drove her away. The mother returned to the spot for the next 48 hours, apparently searching for her missing calf. A different cow uttered loud whining moans for ten minutes after her calf had been shot. Not surprisingly, moose are more than a mere bag of hormones.

Just as in other mammals, moose mothers differ in what they know and how they care for their babies. Some moms are highly attentive to offspring, others laissez-faire. Some stay close, nursing often, while others are seemingly uncaring. Such variation offered Kevin White opportunities to examine the attentiveness of mothers to their calves.

Kevin predicted that mothers would feed more efficiently when calves were inactive. He reasoned that active animals would be more prone to predation than inactive ones, a supposition based on research comparing the survival of sedentary versus active kangaroo rats. Among these large hopping mice with big ears, those that moved around more lost the protection of their below-ground nests and were more likely to become food. Somewhat similar findings stem from observations of porcupines, skunks, and kangaroos. The more mobile die more frequently, though in these cases, it was due to roadkill.

Kevin's curiosity about vulnerability to predation and strategies of maternal protection led him one step further. He asked whether mothers varied in attentiveness when their calves were sequestered in vegetation or feeding. The mothers were more wary when calves were up. In other words, mothers were sensitive to the activities of their young. Alaskan mothers were pensive and aware of their environs.

Mothers might also influence the behavior of their young. Because of our interest in mechanisms of learning and culture, Kevin, Ward Testa, and I asked another question: whether youngsters developed antipredator behavior similar to that of their mothers. To find out we contrasted vigilance between solitary yearlings and those still with their mothers. Attentiveness did not differ, but solitary yearlings were more variable. They were also preyed upon more frequently than yearlings who continued to associate with their mothers. The increased vulnerability of solitary yearlings may have been a simple consequence of their smaller size and the lack of protection afforded by mothers instead of the variability in their behavior.

Rather than asking what young derive from being with their mothers, I posed a slightly different question: during danger, do young remain with their mothers or do they flee independently? The idea cuts to an understanding of subsequent behaviors that young may acquire from their mothers.

Moose of all sizes and shapes must cope not only with grizzly bears and wolves but also with human predators. In the region in and around Jackson Hole, about five hundred animals—mostly males—were harvested annually between 1972 and 1992—more than ten thousand in all. However, because females were also harvested, non-park animals might be more wary of people than the non-hunted park animals. Indeed, moose from the two regions respond slightly differently.

I reasoned that if calves develop behavioral responses to predators through associations with their mothers, then both cow and calf in protected regions should differ from moose living in areas where they are hunted. To address the question—do young remain with mothers?—I gathered data on the frequency with which eight- to ten-month-old calves remained with their mothers when confronted by humans. In this case, the hunters were Tom Roffe and myself during our collaring operations. The contrasts were between mothers with calves in the park and those up to sixty miles beyond it.

Inside the park, we darted more than 35 mothers. In all but one case, the youngsters remained with their immobilized mothers. Seven times, the calves offered rudimentary protection. Although these were not direct attacks on us, the calves vacillated between aggression and curiosity in their approaches. Their ears were dropped and the fur on their necks stood up. Outside the park, calf behavior was dramatically different. Only once did a calf remain, and he never assumed an aggressive stance.

The behavioral differences between areas were fascinating and carried implications for understanding how mothers might shape the behavior of their offspring. Where mothers remained and displayed a lack of concern to humans, calves too showed little or no trepidation. By contrast, calves living in areas with human hunters did not wait around for their immobile mothers to rise. Instead, they fled, whether or not their mothers were with them. Their flight behavior was indicative of fear. Had calves been capitalizing on the behavior of their mothers to learn about predators, then it appears that moose from different regions reflected cultural divergence.

APPROACHING ANIMALS IN the wild is always tricky. Habituation is a behavior that human hunters have long capitalized upon to seize their prey. From cave paintings in northern Nigeria, it is known that the Hausa people disguised themselves with headdresses made of hornbills. In doing so, they capitalized on prey habituated to the ground-dwelling birds to close the gap and harvest unwary mammals with their arrows. Many clever disguises have also facilitated close views. Ostrich and zebra costumes have helped biologists and filmmakers get in tight with their subjects. The skins of wolf, bison, deer and other animals have also been used.

To conduct olfactory experiments on moose, I had to be within forty yards of the animals. At that distance, my delivery of urine-stained and other snowballs was generally precise. Since most of the females I studied were nonhunted, approaches were not too difficult, especially when I moved slowly. This, however, was not always the case.

One day I asked a costume designer from the set of the first *Star Wars* movie to make an antler-less moose suit. She agreed. Fabricated from dark cloth and Styrofoam, complete with a cape, large ears, and a long snout, it cost $250. The four or five times I put it to use, it worked well.

One of its major benefits came to light when the media discovered it. Segments ran on local television, in major newspapers, and in conservation and wildlife magazines. The tabloids also loved it, and an account even appeared in *News of the Weird*, and a friend, riding one of London's trains, reported that the story had crossed the Atlantic. A one-hour special on moose and their relationships with bears and wolves was featured on *Animal Planet*. Fortunately, the moose masquerade was not central to the story.

I was pretty sure the costume held more comic than conservation value, so when *The Late Show with David Letterman* invited me to appear, I agreed, but only if they would honor certain conditions. The most important was that they feature thirty seconds of serious footage from a past project to establish my credentials as a scientist rather than a buffoon. They did not agree. I did not appear.

LUCK HAD BEEN on my side. Data now existed on variation in juvenile survival from two regions within Grand Teton Park, one where wolves had effects and one where they did not. I had also accumulated information on the behavior of mothers and calves where human predators were absent. My zeal for understanding how predation affected behavior across different regions and across time swelled. As I checked my e-mails, however, my enthusiasm crashed.

> Joel, report back to campus. You are to terminate your research immediately. As a respected member of our faculty, we need you to exercise leadership now. Be here next week. —Bernard

The words "terminate" and "immediately" seared in my mind. I had no clue how to respond. Was I really to end my research because the Dean of the College of Agriculture at the University of Nevada–Reno demanded so?

Trained in beef production in Kentucky and built like a bull, the dean had power and standing far beyond academic circles. He charted the path for agricultural research in Nevada, maintained a tight rein on faculty, and had been highly supportive of my tenure and promotion to full professor ten years earlier. Although some felt the dean had little tact, he was an ardent champion of research, particularly mine. Whereas some faculty moaned that my studies had little to do with Nevada or the fact that my teaching was collapsed into a single semester, the dean had been dismissive of such complaints. He maintained that a faculty's strength lay in its diversity.

Among college deans in the American West, those from land grant universities carry disproportionate strength when it comes to setting research agendas. Rarely is there a focus on wildlife conservation. Nevada was an exception, as their College of Agriculture had programs aimed at spotted owls and shorebirds. Words like "conservation" were not often part of other aggie curriculums.

An interesting and historically-sculpted ideology prevailed; predators caused most problems. The decline of the western sheep industry was still blamed on them, and millions of federal dollars were spent annually killing the likes of coyotes, cougars, bobcats, skunks, and bears. Remarkably, the sheep industry in the eastern United States had also collapsed. Since coyotes and cougars were either not even present or were at such precariously low densities, they could not have forced the downward spiral. Market forces were the actual cause, but that was rarely taught.

The predator bias was so strong that it reached senior levels. This was evident from the first question posed to Brian Miller, a candidate for a faculty position in New Mexico State University's College of Agriculture. The dean asked, "Do you think we should reintroduce grizzly bears?"

Before Brian mustered an answer, came the next one-liner, "I don't like grizzly bears."

A similar philosophy was evident at lower levels. The head of the Department of Fishery and Wildlife Sciences said " 'endangered species' is almost like a bad word around here."

Land grant institutions and the public have always had a tenuous relationship between societal needs and research interests. At Washington State University, funding was approved for a pesticide lab but not a center for sustaining agriculture and natural resources. In Wyoming, the agriculturally dominant legislature asked for the termination of a tenured professor, because her book suggested that traditional livestock grazing had impeded the conservation of biodiversity.

Whatever it was that threatened tradition in these western land grant universities was serious, but it paled in comparison to the violence and damage caused by the lawless fringe groups known more appropriately as ecoterrorists. The University of Washington's horticulture lab, which housed genetically modified poplar trees, was damaged by an incendiary device causing $3 million in harm. The animal facility at the University of California, Davis, sustained $3.5 million in deliberate damage, while Michigan State University's animal laboratory was firebombed.

My brouhaha at the University of Nevada paled by comparison, although I still wondered why my immediate return to campus was so urgent. My research support came from federal grants that I had procured and that the university administered on my behalf. In the end, I was told that I could no longer leave the state, because state funds paid my salary. The logic did not quite make sense. Other faculty visited laboratories or field sites out of state, and even the dean conducted university business in Washington, DC. Fortunately, higher administrative officials at the university intervened, and I was assigned to the graduate school for research. My projects would continue unabated.

While the outcome delighted me, there was a personal toll. My keenness to work with a budding new generation of students and novel ideas continued. I was happy for excellent colleagues, but I decided to keep my eyes open for a better professional option elsewhere. Additional seminars about mice on treadmills or tomato flavor genes were unlikely to diminish my growing alarm about the pace of environmental challenges in Nevada and well beyond.

IN 1973, A research station that operated in the Tetons ended its 25-year presence. Run by the New York Zoological Society, its closing coincided with my first trip hitchhiking through Jackson Hole. The Bronx Zoo operated the facility. Little did I appreciate how entwined our paths would one day become.

Founded in 1895 but now called the Wildlife Conservation Society (WCS), this zoo-based organization currently operates three hundred projects in fifty countries. Past efforts have ranged from empowerment of indigenous people in Africa to the study of migratory whales in Patagonia. African elephants, Asiatic gazelles, and Brazilian jaguars have all been targets of major efforts. Although WCS has always had a presence in North America, after the closing of the station in Jackson Hole, its initiatives concentrated on research, protection, and the establishment of reserves in Asia, Africa, and Latin America. Yet it was something that occurred in Rwanda that would lead WCS back to North America, into Wyoming, and eventually to me.

Two Americans, Bill Weber and Amy Vedder, who had worked out of central Africa on mountain gorillas, returned to the Bronx. They recognized that we, as Americans, proffered advice about conservation elsewhere but were unwilling to live with grizzly bears or wolves in our own backyards. Bill suggested that WCS reinvest in North America, particularly in Alaska, Yellowstone, and the Adirondacks.

WCS had already established a relationship with me, supporting my earlier work on rhinos. Over the next few years, WCS helped fund my efforts in Alaska and in southern Yellowstone. I was anxious to know Bill and his organization better, so I flew to New York. WCS scientists asked about my research and other activities. Unlike in academic circles, there was no need to justify why conservation was relevant. It just was.

A few years passed, and then one spring Bill visited my study area. As we drove past the glistening Tetons, I was frank.

"Bill, why don't you just hire me? You know my work. You know how I work."

"Let's see. You currently live up in Sierras. You conduct studies out here in the Yellowstone system. Would you seriously come back to the Bronx?"

"The Bronx? Um. No. Isn't it WCS that always maintains that a field presence is essential for conservation? I want to be based here."

Bill paused and looked down. He rotated his head, absorbing unfettered vistas in every direction. "You mean you expect us to pay you to live out here, while I commute back and forth between Westchester County and the Bronx?"

Our discussion ended.

chapter 8

A Tiger East of the Sun

[We] stood . . . silently, a few minutes in the hope that some sound would betray the presence of the tiger, but there was silence. . . . I felt misery and fear.

VLADIMIR ARSENIEV (1902)

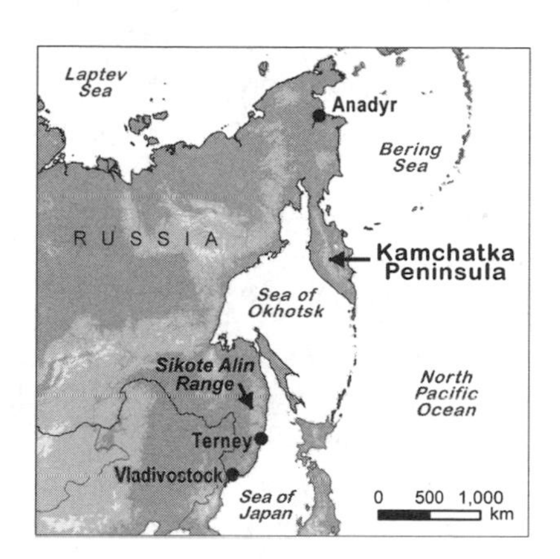

FORESTS OF THE Siberian Far East are thick with rhododendron, oak, and ginseng, Korean pine and spruce, and even Japanese yew. Thickets are profuse, and the danger from predators is palpable. One hundred years ago, Russian geographer Vladimir Arseniev explored the region with an indigenous hunter named Dersu Uzala. Of Tungus-Manchu origin, Dersu had been attacked by *amba*, meaning tiger in the local dialect. Dersu knew how to survive. Arseniev had a good guide and friend.

This area is frozen in time, a place where the instability of Pleistocene glaciers never touched the soil. The faunas and floras have morphed from wildly different zones—subtropical Asia, the Himalayan region, and the boreal forest. In this area are crested hoopoes and imperial eagles, hawk owls and broad-billed rollers, leopards and caribou. This is the Russian Far East, the northernmost home of the tiger. These specialized carnivores stride across white sands lining the Sea of Japan. They consume seal, and they live in deep snow where temperatures plummet to –40°F.

Tigers eat a huge amount, and their diet is varied. On the Indian Subcontinent, it may be wild pigs, an occasional macaque, or any number of Asiatic deer—sambar, chital, muntjac, hog, or axis. Tigers have been known to feast on native cattle, such as banteng and the almost bluish gaur, each weighing up to two thousand pounds, on tapirs on the Malay

Archipelago, and even a shy and beautiful dark antelope, the kouprey. In western Nepal, the calves of greater one-horned rhinos are savored.

In the Russian Far East, tiger diets are unique. Large wild cattle are replaced by smaller deer, such as the musk, whose males have protruding canines. They also eat the roe and sika deer. Wild boar become key players. As their abundance grows because of their association with oak forests, they increasingly become tiger bait.

In chillier realms are two species of large deer, elk, and moose. These are the same species of elk and moose consumed by grizzly bears and wolves throughout Asia and Europe, and they differ little from the ones hunted, eaten, watched, and revered in North America. Pigs and elk together account for more than 85 percent of the tiger's diet, which also includes such anomalies as the regal white-tailed sea eagle. Their remains found at carcasses have testified to a feathery end.

While popularly known as the Siberian tiger, the felid name is more appropriately the Amur tiger, named for the Amur River Basin. This sparsely settled region, twice the size of France, drains the Primorski and Khabarovsk krais south of the Siberian Shield. Confusion about nomenclature does not stop with tigers.

In the Russian Far East, as throughout Europe, elk are known as red deer. In Russian they are called *ilch* or *izubar.* Large bruins with a dish-shaped face and a well-defined hump are grizzly bears to Canadians and Americans but brown bears to everyone else. In Russian, the word is *medveeyet.* Like elk and brown bears, moose, too, are Holarctic in distribution. The same species occurs from Mongolia and Manchuria to Europe and throughout boreal North America. In Europe and Scandinavia, moose are called "elk." In Russia they are moose, the local word being *los.*

It was the *ilch* (elk) and *los* (moose) that lured me to the Sikhote-Alin Mountains, an area once hunted by the Chinese, by the Ainu of Japan, and more recently, by Dersu Uzala. Today, some indigenous Udege remain. But the most common hunters of this six-hundred-mile-long mountainous chain are Russian.

Along the southern end of the Sikhote-Alin, just north of the Korean border, are numerous carnivores. Amur leopards, like tigers, make a living in the cold and snow, although the population hangs by a thread. Dholes have been less fortunate. These fleet, pack-living, hunting wild dogs with ecological roles similar to wolves or Cape hunting dogs disappeared thirty years ago. Other carnivores are doing better. The masked raccoon dog is a small canid no larger than a big house cat. Its diet consists mostly of insects and small mammals. On the mainland, they are rumored to hibernate, although they are active year-round in nearby Japan. Asiatic black bears are abundant, feisty, and aggressive. If lucky, one can also glimpse lynx, yellow-throated marten, Siberian polecats, and wolverines in the boreal forest.

MY SLEEPING HABITS had gone haywire in anticipation of my forthcoming field efforts in the Russia outback. I thought of Michio Hoshino, the acclaimed Japanese photographer whose stunning pictures of Alaskan bears and moose brought global recognition. Michio's partially consumed body outside his tent in the Far East and the stories told of drunken Russian pilots offered ghastly reminders of the perils of brown bears and Aeroflot crashes. Thoughts of danger and data kept me stirring.

My recurrent nightmare involved Manchurian moose. I chased them through an unfamiliar countryside, but just before catching a glimpse I would jolt awake and lie sleepless for hours. Other biologists may be riveted by dreams of scintillating polymers, RNA sequences, or some truly invigorating experiment. Not me. I pursued the moose of my dreams with good-natured local people, all short and squat. They wielded machetes and chopped down the impenetrable vegetation. Then, the bulldozers arrived and the forests were vaporized. I became happy because I could see. I could watch animals. I could get data. The irony was that in my imaginary pursuits, I would do whatever was necessary, even annihilating the very environment that the animals so depended upon.

Russian infrastructure and attitudes offered their own challenges. I was told that many think Americans are wimps—fat, soft, and whiny. My collaborator in the Russian Far East, John Goodrich, suggested that I be in good shape. Tall and sinewy, blonde and bearded, John had been studying tigers for ten years. He now worked in WCS's Asia Program. A former graduate student of mine, John encouraged my research visit. I was fortunate, for John now lived, breathed, and spoke fluent Russian.

To avoid humiliating John—actually myself—I strapped on my pack, added twenty pounds, and began trail running, up and down trails in the Tetons. Over eight weeks, I'd alternate my daily routine, jogging with that pack one day and walking as fast as possible with a heavier forty-pound version the next. But even after falling into bed physically exhausted, restful sleep eluded me. Apprehensions remained.

The terrain would be unfamiliar. Would I find moose? John had never seen *los*. Even if I found them, could I gather data? There were other issues. How did Dersu, Arseniev, and today's Russians avoid tigers and bears while they slept? What mistake, if any, did Michio make that caused him to be dragged from his tent? Should I return to my primate ancestry and sleep in trees? How would I transport my heavy gear to the remote Kolumbe River Basin on the Sikhote-Alin's western slope?

There would also be the human dimension—Russian scientists and rural residents. Cloistered and subjugated for years, would they be helpful or distant, cooperative or uncaring? Had the recent collapse of the former Soviet state and the advent of Perestroika affected this wild side of a country as it had Moscow, eight time zones removed?

THE CUSTOMS OFFICER in Vladivostok was focused on one thing. Me. This was the fourth time an official had asked for my passport. Then I was whisked into a tiny, windowless room.

Two uniformed police arrived, then a third in military garb. We sat, swathed in dark silence. The walls were drab gray, the floor cracked. The only picture I could see was in my mind. Sonja shouted "bye bye daddy!" and blew kisses as I boarded the plane in Jackson Hole. It headed to Anchorage and then across the Aleutians before landing on the dirt runway on the Anadyr Peninsula. After refueling, we stopped in Magadan, and then on to the Chinese border where we deplaned in Khabarovsk. Now I sat in a virtual dungeon with two policemen and a soldier.

The door swung open and slammed closed. The uniformed woman matched my stereotype perfectly—matronly, weighty, and insistent. She tried speaking English but I understood nothing. The police gestured to open my trunks. Instead, I kept pointing to the visa and holding the signed letter from a Russian institution that invited me for research. It was no use. The inspectors had found my playback gear. They next tore into my trunks, duffel, and backpack.

I worried about their reaction to my canister of pepper spray and my stun gun. The former was for carnivores, the latter for people. The inspector held the stun gun, cranking his head inquisitively as if a puppy discovering a first bone.

The taller man with a scarred cheek smiled, his missing teeth accentuating the silver caps of the remainders. He grabbed the gun and raised it high above his head, a scene reminiscent of an athlete's victory celebration. Horrified, I could see that he was about to engage the trigger. I protested. He did not understand. The room fell quiet. Blue electrical waves pierced the deadly silence as an arc crackled across two wires. Depressed, I sank into my wooden chair. *This is it.* All eyes focused on the two electrodes. But the time to wonder what Scarface would do next had already evaporated.

He grabbed the solider, pressing the electrodes against his exposed

PLATE 1. Female moose play crucial roles in imparting knowledge.

PLATE 2. Adult male moose toward evening—with Tetons.

PLATE 3. The tenderness of birth.

PLATE 4. Wolf with moose leg bones in Alaska.

PLATE 5. Radio tracking moose from the air.

PLATE 6. Sonja waits with Sage for her early morning ride to school.

PLATE 7. Glaciers and peaks of Wrangell–St. Elias, Alaska.

PLATE 8. Wolf (left) approaches two grizzly bears in the Talkeetnas. (Photo by J. Lee.)

PLATE 9. Greenland and an open fjord in March.

PLATE 10. Musk ox on a ridgeline overlooking the frozen ocean.

PLATE 11. Polar bear and pressure ridges, Svalbard. (Photo by F. Camenzind.)

PLATE 12. Solitary reindeer on Svalbard.

PLATE 13. Snow leopard.

PLATE 14. Tiger near the Sea of Japan in the Russian Far East. (Photo by J. Goodrich.)

PLATE 15. Khulons in the Gobi Desert.

PLATE 16. Ibex in snowstorm.

throat. His finger was on the trigger. *Was this play or for real?* My eyes opened wider. *Fuck—I'll never get out of here. The gulag will be next. I'll never get to see* los. The other cop intervened. The stun gun was released. Everyone laughed.

I had little time to breathe as the uniformed babushka insisted I accompany her into another room, leaving all my bags just as they were—open. Her English was now perfectly clear—$250. I agreed. The price of freedom was cheap.

Outside, the air was languid, humid, salty. A recent typhoon had battered the Sea of Japan. Still nagging was the question whether my supposed guide, Olga Zaumyslova, who lived about 350 miles north, would actually be there to help me find moose. Olga had grown up near Gorky Park outside Moscow. Her great grandfather had been killed by a moose. Olga had fled westernized Russia years before for romance—a love for wild places and big cats. Although a tiger had been seen within a couple hundred miles of Moscow, it predated Olga's birth by two hundred years. Elsewhere, other striking species remained. Cheetahs persisted in Turkmenistan in the former Soviet states, until at least 1962. Flamingoes and striped hyenas still do.

A single mother with an eight-year-old, Olga was educated as an animal psychologist at Moscow State University. She also had an intense interest in *los* and maintained records of all moose sightings since she moved to the Primorski Krai. Old or young, fit or babushka, English-speaking or not, I knew very little about this mysterious woman. What I did know was that she would be pivotal in my attempt to find moose.

My work should, I thought, be straightforward, given the data I had already gathered. In Russia, I only needed to locate my quarry, remain stealthy, and simulate sounds. My goal was to determine whether these Asian moose differed from those in Alaska and the Greater Yellowstone Ecosystem. Because moose in this part of Russia had three dangerous predators—tigers, brown bears, and wolves—I predicted they would feed less than those from elsewhere. By adopting this sort of comparative approach, I hoped to determine whether moose that were currently naive would differ from the Asian ones that I expected to be savvy to tigers. If North American moose were still responsive to tigers but not to other unfamiliar but presumably dangerous predators, such behavior could indicate a response steeped in their history.

In Alaska, the true king of beasts had lived—*Panthera leo atrox*, the North American lion. This cold-adapted, thousand-pound meat eater ranged from the open steppe and Arctic tundra to the Los Angeles Basin twelve thousand years ago. More than seventy died in the La Brea tar traps alone. The skeletons of these large cats differ little from modern African lions or Asian tigers other than being bigger. I planned to use my tiger recordings to gauge moose responses.

My approach to the *ilch* (elk) was similarly motivated. They had once lived side by side with a variety of different nonhuman predators. Now, however, North America had only a few dangerous predators left—wolves, cougars, grizzly and black bears, and humans.

In this non-glaciated region of eastern Russia, I'd observe how both *ilch* and *los* (moose) respond to three worthy meat eaters: tigers, bears, and wolves. I knew moose might be more difficult to study here because of their propensity to inhabit deep forest. By contrast, throughout their Holarctic distribution, *ilch* enjoy open habitats where grasses are the preferred fodder. If that were also true here, my success might be better in open glades where I would not only broadcast sounds but observe responses.

I hoped that Rudyard Kipling's 1894 accounts of Shere Kan, the tiger in *The Jungle Book*, were wrong. According to Father Wolf, when the tiger arrived, "He will frighten every head of game within ten miles." If this were true, I might not have a chance to do my playback experiments. Perhaps *ilch*, unlike *los*, could be watched from a long way away.

Other areas might also have proved useful for understanding how the loss and addition of predators shaped landscapes and attendant prey. Tigers once inhabited both Japan and Sakhalin Island to its northwest. Although excellent swimmers that have navigated at least nine miles by sea, tigers undoubtedly colonized these island chains by crossing land bridges from mainland Asia some eight thousand years ago.

The question of interest was not whether tigers instilled fear, but how long after their extirpation until prey would forget them. Beyond Aldo Leopold's 1922 tale of deer sniffing for extinct jaguars, the Bisaya of Borneo maintain memory of tigers, claiming their ancestors hunted them only a few centuries earlier. So it comes as no surprise that prey might also have memory of tigers. Not only did Japan have deer that now live without tigers, but on the northern end of Russian's Sakhalin Island, which stretches to the frigid and sometimes ice-bound Okhotsk Sea, there were also deer with no recent exposure to tigers.

If I had wanted to study other prey that currently lived in recently predator-devoid regions, I might have selected Japan. Although brown bears still live on the northern island of Hokkaido, wolves became extinct across the entire archipelago only a hundred years ago. Wolves had been tolerated until the fall of the feudal government of Tokugawa Ieyasu in 1868. Soon thereafter, Edwin Dun, a rancher from Ohio, visited Japan and imported American attitudes of hatred. He also introduced strychnine so that poison could do the killing. The last wolf perished in 1905. As a result, numerous prey species now live mostly in fear of man.

One is the Japanese deer. With tigers and wolves extinct, Japan still has bears. The Asiatic black bear is aggressive and widespread. It's also likely to be more omnivorous than carnivorous. Whether Japanese deer respond to past predators is uncertain but Japanese scientists have capitalized on experiments to find out.

In Wakayama in western Japan, as throughout the United States, deer have been involved in traffic collisions. In Japan, the more serious accidents have involved trains. Tiger dung was once used a couple of years ago in an attempt to curb the deer's interest in railroad crossings. It did not work.

THREE HUNDRED AND fifty miles separated me from the customs agents at the airport. I had gone north to Terney. Hemmed between the Sikhote-Alin and the Sea of Japan, the unpaved streets of this quaint village were a pleasant change from the dangers of Vladivostok. Instead of traffic and police, soldiers in camouflage and refugees from Chechnya, the town's dirt streets were filled with cattle and goats, chickens and dogs.

My hosts, John Goodrich, the American tiger biologist, and Linda Kerley, lived in a simple house at the top of a hill, overlooking the bluish waters of a crescent bay. Behind the house were woods that stretched into broad leaved and conifer coated hills. Their office was simple—a small dining room table. Pictures of tigers and friends adorned the walls. A detached pit toilet sat thirty feet from the house. The lack of heat would challenge bodily functions during Siberia's legendary winters. In a separate building was the banya: a stove, pile of wood, and a pail for water—the shower.

Nearby was a garden in which John and Linda tended their winter sustenance—potatoes. John elected to spend days harvesting them rather than working on tigers, analyzing data, or training local people. When he saw my perplexed look, he explained that to maintain respect, he had to perform the labor. It was the Russian way and he had to fit in. Harvesting local game and holding one's own with vodka were part of the plan.

To fit here in the East, the westernized approach would be ignored. Helping and working collectively would be the only way to act if these Americans were to blend in. Like John, Linda had to work to establish herself. Despite her PhD in biology, she, too, worked in the garden. She also radio collared bears and tigers and tracked them from the air, something that gained respect from the Russian militia. Using their ancient Soviet helicopters to find tigers, they flew Linda along the deep clefts of the Sikhote-Alin. She learned the language, interacted at the market, understood the wild landscape, and was as committed to the tigers as to the Terney way of life.

OLGA, ONCE DUBBED "the quiet woman of the forest," was up in the mountains. Since she was the only person who knew the area with moose, the distant Kolumbe River Basin, I would wait for her. The opportunity to be idle offered a chance for reconnaissance.

Each morning, I jogged on muddied roads past dogs and cattle. The local people and constable averted their eyes or gazed as if frozen into the Siberian landscape. I smiled or nodded to my temporary neighbors, but the overtures of friendliness were ignored. People whose very existence had been dominated by forces beyond their control and who had little exposure to outsiders were justifiably indifferent. With my long hair pulled tightly into an upright ponytail that bounced with each step, and wearing shorts, I was anything but local. After a week, a few faces that grew familiar to me still stared.

I then learned why. I was rumored to be a holy person. Few from Terney had seen people with long hair; Terneans found me to be an oddity, a spiritual Asian with round eyes.

As days passed, I combed beaches, visited markets, and watched Stellar's sea eagles. I swam in cold lakes and hiked through areas known by name as Blogadatna, Häntä Mi, and Sheptoon, all wild regions that harbored tigers and bears, Daurian redstarts and gray wagtails. A forest guard graciously invited me to tea, where I learned the hard way that the by-product of local forest fungus was not for my stomach. The committed guardians were so poor that tea from China or India was a luxury they could not afford. I also checked out the local airstrip where the military practiced landings and takeoffs in their decrepit Soviet helicopters.

As we prepared to lift off, Linda briefed me on the rules. There was one. Don't breathe the leaking avgas, which was contained within a drum, strapped like a bomb to the floor. Seat belts did not exist. If safety were an issue, I was in the wrong part of the world.

Linda's goal was to find tigers. As the propellers whopped in preparation for takeoff, I covered my eyes to avoid being burned by the enveloping vapor. Above 2,500 feet, colorful forests turned from broad-leaved to coniferous. Larch, some 13 species of birch, and oak gave rise to pine and finally to spruce. Meadows punctuating a deep green canopy were burnt golden by summer's end. There were no roads. As we moved inland from the jagged coastline, a faint ping sounded in the headset. It grew louder. Tiger. I was in heaven.

This one, named Misha after a local Russian researcher, was below. Suddenly the signal faded. We must have passed over; he was in the canyon below.

The whop, whop, whop of the helicopter increased as we dropped down. With no door or harness, I leaned out, the air pummeling my face as I tried to glimpse a striped shadow. It was no use, the vegetation was too thick. An hour later, we touched down. Despite not seeing a tiger, I knew more about the habitat that I would have to hike to find *los*.

With Evgeni, Anatoli, John, and Linda, we tracked tigers on foot, moving silently through eight-foot-tall grass and hazel saplings. We swam naked in gentle surf where months earlier Linda saw an *ilch* feeding on kelp as her new calf learned that sea water was salty. We visited sites where roe and sika deer took their last breath, and a tiger slinked off before we saw it.

My fortune grew even better. *Ilch* appeared in the glades at dusk, both down by the Sea of Japan and up in the mountains. I recorded data on their behavior before and during playbacks of tiger roars and other sounds. We watched some animals at three hundred yards, others from seven hundred. Over several days, I gathered information on about thirty different females. My sense was that they feared the sounds of tigers and left the area immediately. But that was impression only because group sizes differed, the habitat was thick in some places and open in others, and the distances that separated me from the *ilch* also varied. Animals in smaller groups might have felt more threatened; animals in thicker vegetation may have felt more concealed. Until I analyzed the data, I was left with only a feeling, and feelings do not constitute facts.

While I waited for Olga's return, I walked alone in woods and hiked up and down hills. A chance meeting with *amba* was never far from my mind. The first written records of human-killing tigers from the Russian Far East to reach the West were in 1870 by Nikolai Prejevalsky, the explorer whose name is synonymous with Asia's true wild horses. Tigers from Manchuria were also known to have killed railroad workers in the 1920s. In subcontinental Asia, some 425 tigers were slain over a 25 year period, partly because they were considered responsible for the deaths of 186 Indians. By some estimates, more than a million Asians had been killed by tigers during a four-hundred-year span. How credible such estimates were was unknown, but tiger bait was one thing I did not want to become. I guessed that my self-centered view about entry into the food chain was no different than all dwellers of this planet, human or otherwise.

Tigers are serious predators. Before my arrival, a brown bear had been treed, although they are not known to be good climbers. An adult bear had also been killed. To imagine that a four- or five-hundred-pound brown bear might not be the most dangerous badass in the area was more than sobering. If bears could be killed and their offspring not defendable against tigers, what chance did a 150-pound human have?

Sleeping in the forest gave me pause. Not only were bear densities great, but the explorer Arseniev once said that his nerves "were strained to the utmost. Every sound, every rustle started me on the run. The whole place seemed alive with bears." To those sounds, I could add thoughts of bear-eating tigers.

TIGERS ARE IN deep trouble, and predictions of their loss have rung true. Scrapes and spoor are no longer seen in Kazakhstan, Pakistan, or Afghanistan, nor on Java or Bali. What once were Edens with our planet's most elegant predators are now paradises lost. Other regions contain only small remnant populations that continue to be at risk of extirpation. But along with these losses and remaining fragments come future hopes, some kept alive in the plans and bold actions of enlightened governments and conservation organizations. One campaign by the World Wildlife Fund's Eric Dinerstein links protected areas across an arc of Terai grasslands south of the Himalayas—from Bhutan through Nepal and into northern India where tigers can roam with wild oxen, rhinos, and pigs. In India, there are still 2,500 tigers, perhaps 200 in Vietnam, and a few may hang on in North Korea and adjacent China. Due to its sheer size—up to five hundred pounds—*amba* require a good deal of space and healthy prey numbers to remain viable.

Being large carries dual liabilities. Because tigers are meat eaters and occasionally kill people, they generate both economic and human safety concerns. Some argue there is no room for tigers in today's world, that their time has passed. Others prefer to confine them to nature reserves. But when humans reach the point that there is no space left for tigers because they eat the wild ungulates that we covet, how can we seriously believe in a future of our own? If we cannot accommodate tigers, it will be only a few decades longer until we will no longer have room enough for our livestock, and intolerance for our fellow animals will turn to ourselves. In many places it already has.

THE MAN AT the dock knew I must be the American. Weathered, tall, and without a trace of fat on his sixty-year-old frame, Alexander Mezlinincoff spoke little English. The seas were too choppy, according to him, so we would wait another day to look for coastal ungulates.

Three days later, the sea calmed. Again I met Dr. Mezlinincoff, who worked as a biologist at a research institute when not gardening or fishing.

We loaded his little skiff, tossed in an extra can of petrol, and pushed off. Ghoral were supposed to live along the cliffs. This goat-antelope is related to serow from Japan and chamois from the Alps. Spread from northern India and Myanmar to southeastern Siberia, ghoral are not continuously distributed; the next nearest population were far west of Manchuria.

Traveling at a good clip across the glassy ocean, I asked if there were many tigers near Terney. "Da, da," replied Alexander. I understood his "yes," but repeated my question, attempting to force a more thorough answer. I wanted to know how many he had seen. Holding up my hand, I flashed five fingers. He suggested that there were many but could not count them all.

Confusion showed across my wind-blown face. *Tigers were rare—other people, even woodsmen, had spent their entire lives in the bush without seeing ones. How could Alexander have seen too many to count?* I looked at my phrase book and pointed to the term I had been saying in English. He appeared puzzled. Soon the puzzle was solved. He thought I had been asking about *taiga*, not tigers. Taiga is another word for northern boreal forest. With so many trees, no one has ever counted them all.

The sea grew ornery as waves turned to whitecaps. Our little skiff bounced. Several shapes appeared along the base of the grassy cliffs three hundred yards away. Having never seen ghoral, I couldn't be sure and I could not steady my binoculars. Alexander moved the boat closer. The engine failed. He barely noticed. I could now make out dark bodies with heads framed by petite horns. Dappled white fur clothed the throats of all five ghoral, and white stockings ran halfway up their legs.

Then we hit rocks. This time Alexander took note, as he landed partly in the water. I yanked him back to the boat, and we both tried to row. As with the helicopter, safety was a nonissue. This was not America. There were no life vests. As we pushed off, Alexander said in surprising English, "I must not have calculated correctly. The winds were stronger than I thought." I kept us from the rocks and he added petrol. After seeing forty more ghoral, we returned to Terney, but not before the Russian Naval Guard intercepted us.

Alexander had failed to notify authorities of our route. Now, with an American aboard, he was worried. He had no permission. I had neither permission nor passport. Alexander explained the problem was not that he had forgotten to call in our ghoral-seeking voyage. He had no phone so he could not call. His comment struck a chord.

Neither he nor his wife, Ina, could afford a phone, despite the fact that both had PhDs. As we touched shore, I offered my heartiest thanks for a very special day. I did not know whether money was appropriate or offensive but suggested that I would like to contribute to their research, to the cost of petrol, to a dinner, anything. Alexander refused. He thanked me for coming, and then shook my hand. Dignity was at stake, not currency.

LOS **IN THE** Russian Far East are hard to come by. Moose densities in Alaska or Wyoming may approach 2½ animals per square mile, a value that was ten to twenty times greater than here. If I were to find moose in tiger territory, I'd have to be lucky.

Olga and I first met in her empty government office. In her late 30s with dark hair, her body well muscled, and her legs strong, Olga's hazel eyes gleamed. She also cast them down frequently, especially when speaking. "I am worried about your feet. Can you hike twenty or thirty kilometers? I must protect you," she offered ever so sweetly.

While Olga worried about my feet, I worried about my pack. It was going to be heavy, and I needed 12 days of food. I visited a market for a few last items. In a dank bunker that substituted for a store, most shelves were empty. The only fresh produce contained hefty doses of unknown pesticides from China and Korea. But the Snickers bars, that delicious, gooey concoction of chocolate and peanuts, caught my attention. Linda explained that never before had American chocolates been available in Terney. Perfect, I thought. I asked the clerk for a dozen, one for each day of the journey. Linda's mouth fell open.

"JB" she said, "This is not like the States. You can't buy all these, even if you have the money. The Russian way is to purchase only one. Let's leave the rest for everyone else." The roots of the old Soviet state had not changed much. A little for everyone rather than much for a few was an ideology foreign to someone from the States.

Linda and John engaged or advised other projects with their Russian counterparts. One included leopards. The shy cats were at low density. Information was needed on their ranging patterns, yet no one had ever tried to radio collar leopards in the Russian Far East. Several American collaborators suggested using dogs, noting that in the western US hounds are used to tree cougars for subsequent immobilization and collaring. When Russian biologists were asked if the plan would work, they laughed, saying the dogs would be killed. Finally, it came time to test whether Russian wisdom would trump American ingenuity.

American dogs were brought to the southern Sikhote-Alin. They discovered the fresh spoor of a leopard. Baying, the dogs charged along the trail. The leopard had not gone far. Rather than run, it turned on the dogs, leaving one dead. Further attempts were aborted.

OUR JOURNEY ACROSS the Sikhote-Alin Divide began in autumn. My pack was only seventy pounds, but it still weighed more than I wished. In it were a few tricks I would need later to forge relationships with my younger colleagues. In addition to Olga was Losha, a strapping 25-year-old technician. With the physique of a heavyweight boxer, he was able to carry my playback unit along with his own gear.

We followed a game trail used by wild boar and *ilch*. Overgrown with vegetation, I constantly fell behind, shambling over roots and logs. The mountains were big with deep cuts. Fortunately, animals are a lot like humans, and rather than going straight up or down, the gentle paths they developed for their convenience we followed for ours. A dinner of lard mixed with boiled potatoes and gelatinous beef ended the ten-mile day. Leftovers became breakfast.

Once we crossed the divide, I breathed easier. My feet were blistered and shoulders and hips sore, but still functioning. We examined dung, watched crossbills, and investigated the remains of dead animals. Every ninety minutes we stopped to rest. On the third day, I arranged about twenty brown morsels that looked suspiciously like moose feces. As I knelt, I pointed excitedly and said "*Los, los*!"

Olga neared. I picked one up, raising it to my nose. I sniffed it, put it down, and picked up another repeating the procedure. I opened my mouth and popped one inside, confirming the nugget's identity, "this is *los*."

Stunned by my culinary habits, Olga's jaw dropped. I was clearly more demented than either she or Losha reckoned. Eating moose shit was not part of a Russian diet or that of any known biped.

I handed Olga her own moose poop, gesturing that she too should enjoy the latest in field cuisine. How else was she to confirm the identity of our mutual quarry? She did not. I tried again, extending my hand with half a dozen juicy nuggets. Her brow furrowed.

The charade had gone on long enough. I asked Olga to say "chocolate" in Russian. She realized then that it was not moose scat on the ground, only a brown and very edible candy that resembled it. We ate it all. The hoax worked well. I decided I would wait a while for my next surprise. Our personal dynamics were improving, language barrier or no.

We reached our final destination, a small Dr. Zhivago–like cabin situated on a cold clear river. The habitat had changed to semi-boreal. Willows lined the banks, and small pockets of spruce grew nearby. A gray heron flew from the marsh. Cottonwoods grew on an island. We had crossed bogs and disappeared into thickets of tall grasses. There had been tiger scrapes and the actual feces of moose. I thought my search for *los* was ready to begin. Olga had other ideas.

She began to cook, first potatoes, then onions, then carrots. This would be our celebratory meal. The tiny cabin warmed quickly, my thermometer registering 95°F. Smoke filled the quarters, but neither Losha nor Olga paid attention to the heat or to the choking particles.

When Knut Rasmussen crossed the Siberian Arctic on sleds in the 1920s, he reported similar generosity by Chukchi women. Like Olga, they heated their small dwelling by fire. The inefficiently ventilated structure rapidly filled with blinding smoke as walrus meat cooked. Other Chukchi sat naked. It was no wonder Russians thought Americans were wimps. I politely excused myself, preferring the buzzing and biting of monkey flies and mosquitoes outside to the inferno within.

Soon Losha joined me and crafted a fishing rod from willow. Eight minutes later, four flayed graylings were on the table. I chopped wood. Two puncture marks decorated a discarded thermos in a pile of rubble. They were testimony to the power of a hungry brown bear. Antlers of *ilch* dangled proudly from above. A scythe was nearby. It was the perfect remedy for grass removal, known otherwise as insect abatement. With no habitat, there would be fewer mosquitoes. I cut grass, Olga cooked, and Losha prepared grayling.

With dinner, I offered Mexican habaneros for a spicy flavoring. There were no takers. Even before finishing our meal, the vodka appeared. After naps in the blistering cabin, Olga and I hiked to a point where she had previously watched a brown bear trying to kill a moose. Although that hunt ended unsuccessfully, we soon discovered a large corpse concealed by dirt and woody debris. Below the mound were the bones of *los,* some still rich with putrid meat. The victim had been dead for two or three days. For comfort I touched my holster containing pepper spray, reminding myself that bears often bed near their kills. I quickly detached the mandible, and planned to boil the teeth later so that I could determine the age of this Manchurian moose.

My relationship with Olga was to change soon. She wanted to return to the cabin. I told her I could not. I must remain in the forest where I could get data; it was a place she and Losha avoided at night because of bears and tigers. Olga would hear of no such folly. To emphasize the gravity, she repeated her observations of four tigers in the area. Brown bears had twice torn through her tents, and had even been in the cabin.

In her sweetest voice, she tried to convince me "I be your Dersu—you be safe with me, OK. Stay." I had anticipated this sort of conflict and suggested that if I convinced her that I would be alright, and she should allow me to remain. Anxious to learn how this chocolate moose shit–eating American was going to persuade her, Olga eagerly awaited explanation.

I dropped my pack and readied for my second surprise. From a stuff bag, I pulled telescoping poles and a reinforced Gortex plank. I asked if she knew what this was; she shook her head "no." I then yanked out a jumar, half a dozen carabiners, and rope. She asked where I intended to climb since the nearest granite and cliffs were miles away. I said I'd sleep right here. Instead of cliffs, I was going to hang on a platform from branches. I hoped the thirty feet up in a tree would be beyond the reach of tigers and brown bears.

Olga just stared, stupefied. Here was an American, and he wanted to be left alone. Sure, in the remote wilds of the Kolumbe River Basin, a place where bears kill 650-pound moose and tigers kill bears. I was her responsibility. But I had obviously thought a lot about how to survive, something I had done, at least up to this point.

To demonstrate, I arranged the ledge, carabiners and ropes, and pulleys and then hoisted myself up, playing with the jumar. I followed with my remaining gear—my small cook stove, a pot, and canisters of fuel. A small packet of oatmeal would be breakfast, power bars and pumpkin seeds lunch, and dehydrated vegetables dinner. I had a powerful spotlight, a dozen high-grade batteries, my pepper spray, and a bivouac sac for protection from rain. Multiple books were to pass the days. Olga now knew why my pack was heavy and she had no choice but to say "yes."

The next morning, I awoke in my tree house to the bugling call of *ilch*. The rut was underway. After oatmeal and coffee, I went to explore. I walked through the forest, where visibility was but a few feet. If I made noise I would see nothing, so despite the danger, I walked silently. Twice the bushes moved in front of me but the sounds quickly grew faint. One time it was a wild boar, the other a deer. Fresh bear tracks dashed my brashness, and any remaining confidence deflated when I saw those of tiger. I needed no reminder of the dead moose that lay only a few miles away. When I finally returned to my home in the tree I said out loud, "Oh, this is going to be just great fun," knowing days of adrenaline-pumping alertness were ahead.

As I waited for darkness to fall, I recalled disjointed conversations with Olga. Although I had wanted to talk about moose, she was more interested in understanding Americans. Why was long hair fashionable? She didn't believe my explanation that it was to swat flies. She asked why Americans wear necklaces—was it for Jesus? Before I could muster an answer, she switched topics. Why do Americans think Darwin is so wonderful? I suggested her sample was terribly biased—a few biologists—and that most Americans were not that knowledgeable about Darwin's ideas.

We then veered to the Russian invasion of Afghanistan, corruption, and poachers of moose. Olga argued that with more money, Russian peasants could emulate Americans—owning big cars, big houses, and fancy things. In awkward defense of the image we had exported globally by our excesses, I suggested that not all Americans were so into money. "Maybe," she conceded, acknowledging that the biologists she had met on the tiger project were not like the Americans she had seen in Moscow or on television.

I asked if she was happy. She admitted that despite being poor, she had food and a home with heat. It had not always been like this. In her early years in the wild forest of the Far East she was always hungry. Once she displaced a tiger from a recent boar kill so she could have meat. It was a slice of Russian life that no American could comprehend.

BY MIDDAY, MY stomach had started growling. I fired up my stove. Lunch would be the same as it had been for days, a power bar ironically called Tiger's Milk. With 150 calories, 45 of which were fat, and with my noodles, I'd be only slightly hungry afterwards. It was cloudy and windy, so I climbed into my bivouac, closed my eyes, and drifted comfortably off. The sleep was needed, since I'd have to be up all night in case *los* passed by.

I assumed that if I was fearful of every sound when out walking, moose must be deathly afraid. Although moose are large and formidable opponents, having killed wolves and chased away bears, their lives depend on their abilities to avoid predators rather than engage them directly. Whether moose or other prey are consciously afraid is unclear. Animals like baboons and elk release stressor hormones when densities are too high, social pressures too great, or disturbances too unpredictable. But prey species, and presumably predators, are probably unaware of how their behaviors transfer to the lifesaving maneuvers of avoidance or capture. Similarly, neither baboons nor humans understand how their kidneys, lungs, or digestive tracts work. Those with functioning systems and adept behavior are obviously more likely to pass genes on to the next generation. Darwin.

Before nightfall, I enjoyed a quick walk. I kicked moldy tiger turds the color of turquoise under the rays of a setting sun. Behind a clump of Korean pines, trails converged. Wild pig, elk, Asiatic black bear, tiger, brown bear, moose, and musk deer had all crossed this way. Tall reeds surrounded a glade a hundred yards across. I walked on moist tussocks and entered a thicket where water ran powerful and clear. I filled my water bottles. The air was soft and the light gentle. A kingfisher flashed past.

Darkness arrived by 8 PM. My work items were carefully arranged—a heavy torch and extra batteries, a data book, stopwatch, and binoculars. A speaker was readied with tapes of different sounds—crow, owl, tiger, hyena, water, and wind—and an amplifier sat atop my sleeping ledge. The pepper spray was next to my jacket, and I sat on my sleeping bag. A packet of pumpkin seeds, pemmican, and my toothbrush would help keep me alert. Unfortunately, a burst of wind grabbed the seeds, blowing the bag effortlessly into the air. I watched as it finally came to a resting spot on the forest floor. I was so hungry, though, that the five minutes needed to re-arrange paraphernalia before descending was worth the hassle. I lowered myself to the ground, grabbed the seeds, and used my arms to pull myself back up.

I sat motionless. There were no signs of life—yet. I waited, but nothing happened. I waited longer. *It must be midnight.* Only two hours had passed. *If there is this much action, this is going to be a long week.* By 11 and in total darkness, I could barely keep my eyes open. Thirty minutes later, something rustled in bushes and a sucking sound moved through mud. My heart pounded. I ruled out tiger. *Amba* was just too stealthy.

With a red filter over the torch, I aimed it into the forest until a large dim object came in and then out of view. Fifty seconds later, it was back—an antler-less moose, no calf. She browsed, ignoring the primate in the perch. I recorded data on feeding rates. The scene was tranquil. I readied the playback unit. I tried the sound of running water first, something that all animals have heard and do not fear. Immediately, she froze. Ears were alert. She lifted her head and sniffed for clues before returning to her meal. Good news—*at the very least I am not scaring her.* I waited another five minutes.

It was time for the tiger's roar. The deep "aouh, aouh" began at a slow pitch. Immediately there was the deep and rapid suction of footsteps breaking through soggy wetlands. Not only was the moose racing through marsh, but different sounds exploded from behind me. *I wonder what those are?* Although the moose ran, I now knew something about her response. At the same time I recognized that this was but one moose and one response.

I relaxed and reorganized while prepping for more visitors. None came. A distant owl broke the austere silence. My eyes grew heavy. By 3 AM, I was asleep, but I awoke at first light.

The sky was gray. With no more action, I made coffee and read until 10. It was time for more exploration. This time I found six carcasses all within a half-mile of a natural salt lick. Among the skeletons were *ilch,* pig, and sika deer. In mud, a beautiful print of a brown bear was less than a yard away from one of a tiger. Three hundred yards further was the deep impression of an Asiatic black bear. No wonder animals were vigilant here—licks were magnets for mineral-deprived herbivores, and these ungulates were magnets for carnivores.

FOUR MOOSE KILLS, two by tigers and two by brown bears, were in this part of the Kolumbe River Basin. Since Russian researchers had no access to radio telemetry until recently, most of their biological information stemmed from commitment and courage. Olga had, at times, spent weeks in the little cabin despite three feet of snow and temperatures of –30°F. Based on work she helped put together, it looked as if bears killed about four times as many adult moose as wolves. Interestingly, this was not so different findings in Alaska, although little was known of carnivore densities in the Russian Far East, a fact that made strict contrasts with Alaska difficult.

One day while returning to my loft in the forest, I lost my way. It would soon be dark, and being completely uncertain of which direction to go, I climbed a Korean pine, hoping to glimpse the sun before it dropped below the horizon. The canopy was too thick and I was hopelessly confused. Then, my footing gave way, and in desperation I lunged for a branch. It snapped, and I plummeted backwards to the ground. Air exploded from my chest and my head snapped back. Dizzily, I groaned. My breathing returned. I was lucky, nothing broken. The bog cushioned my unplanned acrobatics.

Disoriented and soaked, I moved onward. A drizzle completed my drenching. I needed to move quickly. I deliberately bent vegetation while walking in continuously broader circles hoping to see boot marks from my path. If I found one, I'd go until I saw the next, and the next. Finally, I hit a game trail. The rain grew heavier. The trail grew more familiar until I finally reached my tree house. It was 36°F, the perfect temperature for hypothermia.

In the wee light of morning I made out two more moose, a mother and calf. This time, I played the sounds of crows first. The mother became vigilant but did not run. She continued standing and remained another seven minutes before resuming her feeding. I then played back water, and followed this with other sounds. My improving luck continued, and the next night I encountered a male moose. I took no data on him, choosing instead to simply enjoy the moment. A female *izubar* (elk) and two friends came to drink at 5 AM, but I fell asleep as they fed, awaking later to the sound of a bull bugling far away.

Two days later, the weather caved in. I had now been on my perch for

18 hours straight, mostly in the rain. The bivouac worked well initially, but the seams now leaked, and my sleeping bag was wet. I was tired and grumpy but savored the data I had recorded over the past few days on five different moose. Others had also visited, but I was uncertain whether I had sampled these previously.

If I resampled the same animals I ran the risk of biasing my data by using the profiles of the same individuals more than once. It would be like scoring a surfing competition but not knowing the identities of the participants. Just as one individual might receive a perfect score, and then another and a third, if the individuals could not be told apart, it would be impossible to know whether all surfers had outstanding rides or if the same surfer simply performed on each wave. Similarly, for moose, I had to be certain that I was sampling a portion of the population. But with only five different individuals, the total samples I required to reach any conclusions about the effects of multiple Russian predators on moose would not be happening in this part of the Kolumbe River Basin.

I returned to the cabin and asked Olga where else we might find moose. Densities to the north were also very low. Populations had apparently dropped during the last 25 years. Olga did not know why—whether due to poachers, predators, or forestry practices. I wondered whether these moose that lived at the southern edge of their range might be stressed due to warmer weather, as is apparently the case in areas of Minnesota and Michigan's Upper Peninsula, where ticks seem to play an increasingly debilitating role. More clear than anything was that studying moose in the land of *amba* was not easy.

I would have to look elsewhere for prey or rely on my larger sample for *ilch*. The limited data on *los* were not going to be sufficient. My journey was over.

My grim mood did not improve on the trip back to Terney. The forests were shrouded in a deep mist, the sky gray, and a brisk wind blew from the north. The dirt road was chocked full of holes, and three gas-spewing trucks passed, each with dozens of freshly cut logs from the Siberian countryside. Splinters pelted our windshield.

Multinational companies housed in immaculate offices in Korea and Japan fuel the thirst for these woods while showing little remorse for the denizens of this remarkable landscape. When I asked what hope there was for the protection of the vast forests and their animals, I was told the plans should be left with the Russian people. When I stubbornly pursued this further, I found that the voices of the populace were muffled by powerful or corrupt politicians in cities.

"Capitalism is not the answer; it will destroy us," a peasant told me. I thought of Alaska's magnificent Tongass Forest, also under assault from logging interests, and I wondered how we Americans could continue to tell other countries to lock up their forests and fauna as we contradicted ourselves at home.

My pessimism faded when I recalled the efforts of Al Gore and others who spearheaded cooperative efforts. Democracy had only recently emerged, but Russian and American intelligence agencies enhanced the sharing of data on oceanography, climate, and environmental changes. Musk ox had been sent from Alaska to the Siberian tundra, and Przewalski horses were returned to one of their original homes on the Russian Caucuses. And, now, Americans were helping to fund tiger conservation projects run by Russians.

Back in Terney, I noticed the grease-stained shell of an old Soviet helicopter. *Is this really the one that is to take me to Vladivostok tomorrow? This heap wouldn't really lift off the ground.* I thought of the recent crash of an Aeroflot jet whose pilot was drunk.

For the final dinner, our Russian collaborators prepared a feast. There were potatoes and lettuce from the local garden, and borsht. There were seared pine nuts, fresh mushrooms, and ripe berries—all from the taiga—and homemade wine. The salmon came from local streams. Along with wild boar were unidentifiable chunks of fatty animal protein. When moose was served as a surprise tribute, Olga refused it. She loved *los* live, not on a platter. An argument ensued. I was conflicted but honored to be there. Vodka flowed.

I could no longer see straight by the time I tried to leave so that I could pack for my 7 AM flight, only six hours away. A man with the confident swagger of a military officer offered another toast, one of about ten. He suggested I remain with his comrades until morning. His arms wrapped around me in a gesture of friendship. Another bottle of vodka was opened. At 2 AM, I made one last attempt to depart. Blocked again, my stammering, bleary-eyed friend offered unforgettable words: "The helicopter will wait, I am the pilot."

part iii

A Search for Ice Age Relicts

Large mammals were decimated. In North America, thirty-three genera . . . disappeared in the last 100,000 years (or less) while South America lost even more. . . . Large mammals survived best in Africa. . . . No late Pleistocene families were lost from Asia or Africa to match the New World loss.

PAUL MARTIN (1984)

chapter 9

A Continent of Virgins and Recent Ghosts

Perhaps no human culture can be expected to walk lightly upon an unfamiliar landscape—. . . the level of chipped stone turns us into formidable predators of wildlife that did not co-evolve with us. Animals too big to hide, too naive to run . . . are the most vulnerable.

CONNIE BARLOW (2002)

FIFTEEN THOUSAND YEARS ago, parts of North America were lands of ice and virgins. Glaciers abounded and temperatures were crisp. Although dwarf monkeys the size of marmosets had lived in tropical Nevada forty million years earlier, it was only when the cold Beringea Land Bridge became navigable that an upright hominoid colonized a continent whose denizens knew nothing of human hunters. This was during the late Pleistocene, a period of cataclysm and change, of warming and fauna collapses.

Today's Africa has been called the living Pleistocene. Only there do vast herds of wildebeest and gazelles, zebra and giraffes still sweep across spacious savannas. Only there do large carnivores persist in assemblages as diverse as they had been in the past. Unlike other continents, Africa has retained 85 percent of the large mammals that lived between 11,000 and 50,000 years ago. Contemporary Africa is not alone. As recently as 125 years ago, subtropical Asia rivaled the astounding diversity of its southern neighbor. Between the Indian Ocean and Himalayas were lions and tigers, snow leopards and cheetahs. There were brown, Asiatic black and sun bears. Smaller predators included dholes, leopards, and striped hyenas. India's array of hoofed animals was stunning—four-horned antelope and blackbuck, gaur and banteng, chinkara and nilgai. There were wild asses, elephants, and rhinos. Seven species of deer existed, along with other antelopes. The Ice Age legacy that we now associate only with Africa had survived on two continents until the beginning of the twentieth century.

Unlike the lands where the great apes were born, North America's spectacular megafauna collapsed well before the twentieth century. An estimated 73 percent of the species larger than a hundred pounds vanished between nine thousand and thirteen thousand years ago. The extinct were browsers and grazers—a half dozen species of elephants, llamas and camels, two types of peccary, and three kinds of horses. Disappearing also were the predators and scavengers—the dire wolf, the short-faced bear, the *atrox* lion, the American cheetah.

The scale of loss, coupled with changing climates, reshaped North America's ecology, leaving some niches unfilled and others with different players. The sweet bean pods of mesquite now rot on the ground, whereas they once fed mastodons. Honey locust, with their pods rich in seed and sugar, are protected by thick and thorny outer coats. But these hard-coated fruits that once attracted, and were dispersed by, browsing llamas, camels, and elephants, do so no more. Protective adaptations that may have thwarted past herbivores now lie dormant. The disappearance of big or dangerous animals has played a curious evolutionary role, one that spawned the idea that ghosts still haunt North America.

Pronghorn are the perfect example. They run fifty miles per hour, have huge lungs, and maintain an aerobic capacity unmatched by other hoofed animals. But why should this be? Neither wolves nor cougars are fast enough to catch them.

Pronghorn fleetness may exist for two reasons. The first is simple, to outdistance predators, of which the now-extinct American cheetah may have been the most dangerous. Indeed, the sole eyewitness account of a cheetah-pronghorn interaction suggests pronghorn speed alone may not be sufficient to outrun the now extinct predator of American grasslands. In his 2003 book, *Survival by Hunting,* George Frison wrote, "In the 1930s, I saw a man traveling with a pet cheetah turn it loose to pursue a pronghorn, a young female, until she sailed over a deep ravine that the big cat refused to negotiate. The cheetah's owner rapidly retrieved it and immediately departed that area to avoid the possibility of running afoul of local law enforcement."

The second reason, though related, is more intriguing. Pronghorn do not need to run nearly fifty miles per hour to escape current predators, but they do. They are overbuilt for running, argues John Byers, a professor from Idaho who knows more about these speedsters than anyone alive. Pronghorn and their ancestors, of which one species carried six horns, have been in North America for twenty million years. John argues that what we currently see reflect past adaptations to recently lost Pleistocene predators. With enough time and an absence of swift predators, pronghorn of the future may not be as fast as their predecessors.

Charles Darwin did not call these mysteries of anatomy and ecology "ghost effects," but they may be just that. Darwin, however, recognized their importance noting that "after the lapse of time," structures and function do change. It is this interaction between natural selection and time that is crucial to understand which traits linger as a consequence of the past and how this translates into the behavior we witness today.

Important clues stem from islands where some species of birds are currently flightless. With predation relaxed, flight was more costly than beneficial. As early as 1859, Darwin suggested, "natural selection is continually trying to economize every part of the organization [of a species]. If under changed conditions of life a structure, before useful, becomes less useful, any diminution, however slight, in its development will be seized on by Natural Selection, for it will profit the individual not to have its nutriment wasted in building up a useless structure." What differs between flightless birds and pronghorn speed is that the loss of dangerous predators has been so comparatively recent that pronghorn still react as if the ghosts are real. Flightless birds do not.

Islands offer other poignant examples. Cacti occur throughout the world, many with barbed spines and dense protruding masses that generally offer effective deterrents to browsers intent on a fleshy meal. Like islands birds, however, plants, too, can have relaxed defenses.

The vegetation on the Channel Islands just off the southern California coastline is illustrative. Domestic sheep prospered after their introduction in the 1800s, and plant communities today still reflect the destruction. It is the difference, however, between plant forms on islands and the adjacent mainland that is instrumental to the understanding of evolutionary defense. Mahogany, cherry, and bush poppy on the islands all had larger leaves—the juicy targets of sheep—and fewer spines per leaf than did comparable plants on the mainland where native herbivores had prospered. In other words, antipredator resistance had been relaxed on the island.

Forces involving the isolation from predators, time, and the intensity of predation have also shaped behavior, but the challenge has been to sort out the relative role of the past from the current. One island where this difficulty has been particularly vexing is Madagascar. Two striking lemurs—Verreaux's sifaka, with its creamy white fur, a dark brown crown, and white-tufted ears, and the ringed-tailed lemur, with its conspicuous bushy-colored black and white tail—both become alarmed when large raptors fly overhead. This behavior has been interpreted as a reaction to a large eagle that that has been extinct for four thousand years, an inference that is not unreasonable, since large eagles are effective predators of species the size of these lemurs or larger. Coupled with the loss of eagles is the fact that the only current raptors are kites and hawks, which are smaller than the extinct birds. Neither of these is known to take sifakas or ringed-tailed lemurs. The problem with the explanation that the alarm behavior of sifakas and lemurs stems from past predation is that not much is known of patterns of current predation on their young or the level of predation necessary to maintain lemur alarm behaviors.

To further muddy the water, immediate behavioral responses to large overhead birds can be clouded by immediate experience, an issue far from confined to mammals on islands. Both bighorn sheep and pronghorn show intense vigilance after ravens have flown above. There is no evidence that ravens are predators of the young of these species, but golden eagles are. Together, eagles and ravens can lead to massively confusing antipredator responses.

Several winters ago, Jon Beckmann and I found a young sixty-pound pronghorn female that had just been killed by a golden eagle in western Wyoming. Its group mates, numbering more than 770 animals, moved off about two hundred yards. Shortly thereafter, ravens vocalized from above. The entire gathering of pronghorn panicked, running more than 3½ miles and then milling about for almost an hour before returning to feeding. Such mass hysteria cannot be common, but it demonstrates an unusual chain of behaviors set in motion; in this case by the ravens (although the predetermining factor was likely the eagle predation event). Trying to work backward in time to establish the cause of behavior, as in the case for lemurs, will require a better current understanding of prey and predator relationships.

While the bones of animals offer insights about their history as a species and their current or past adaptations, behavior varies across time and space. This has been obvious when examining how isolated populations, like those on islands, cope with predators. Whether behavioral innocence has contributed to the past extinctions has been the subject of speculation.

Pulitzer Prize–winning writer Jared Diamond remarked, "Just as modern humans walked up to unafraid dodos and island seals and killed them, prehistoric humans presumably walked up to unafraid moas and giant lemurs and killed them." In contrast, Diamond also argued that "most big mammals of Africa and Eurasia survived into modern times because they had coevolved with protohumans for hundreds of thousand or millions of years. They thereby enjoyed ample time to evolve a fear of humans, as our ancestors' initially poor hunting skills slowly improved." By combining the arguments of Diamond and Darwin, Tim Flannery, an Australian paleo-ecologist, followed up on the interaction between past and present behavior and predators.

> All wild creatures need to learn what is dangerous and how best to avoid it. Creatures that flee too readily are disadvantaged because of the vast amount of energy they waste and the disruption this brings to their lives. The behaviours animals use to avoid predators are both genetically based and learned. The genetic component is acquired through natural selection and so can only be slowly developed. This may account in part for the fact that most of the world's surviving large mammals live in Africa, for it was there that humanity evolved, and it was only there that animals had the time to acquire the genetically based behaviours that allowed them to cope with the new predators.

If North American and other prey treated humans as if they were unknowns, and had the prey not been preadapted to the arrival of a novel carnivore, then it is understandable that extinctions were so rapid. The prey simply failed to recognize the inimical nature of the new beast, and died swiftly.

Critics of the idea that humans drove the blitzkrieg at the end of North America's Pleistocene claim that the extinctions cease abruptly 8,500 to 9,000 years ago. If prey were so naive, how could they suddenly become so savvy? And, if they did become savvy, what led to the rapid gain in knowledge?

Answers are hard to come by. Defenseless prey experience rapid extinction, but the central issue concerns survival, not death—specifically, whether prey can reverse an onslaught so quickly that they survive. If so, it is germane to ask what mechanism(s) favor survival. Rapid learning must play some role, but detailing what it may be, how it operates, and how it operated in our past are all challenges that have yet to be overcome.

THE IDEA THAT naive individuals experienced heightened vulnerability to death and that populations or species subsequently collapse has appeal. Orcas along the South African Coast eating naive jackass penguins, wild dogs approaching lions to feed on carcasses only to be killed in the process, and Indian wolves grabbing blackbuck are all contemporary examples of prey with little experience encountering native predators. Beyond these are cases where native predators expand their ranges and enjoy success with prey—such as coyotes moving into Quebec and eating caribou calves or cougars dining on bighorn sheep in the Mohave and Great Basin Deserts.

As interesting as such cases may be, few direct insights emerge about the plausibility of prey naiveté as a cause of continent-wide collapses at the hands of human hunters. What might prove more telling is whether African big game behave differently to hunting humans than big game elsewhere, an idea that faces one mammoth problem.

Nowhere in the world today are methods of hunting the same as they were ten thousand—let alone one hundred thousand—years ago. Factors that affect prey responses to predators include far more than time of exposure to prehistoric bipedal hunters. While a few !Kung in the Kalahari still harvest game with traditional bow and arrow, other primitive weapons, such as pitfall traps, snares, and spears, are constructed with materials today that were unavailable in the distant past. Such modern amenities undoubtedly enhance the efficacy of harvest.

We can't know for sure whether African big game differ in predator-savvy from big game elsewhere. The sources of variation in how the wildlife might respond are many. Densities of native predators, past human hunting and nonhuman predation pressures, as well as habitat conditions, may all play roles. So might group sizes, presence of young, the season, snow and wind conditions, heat, health, and habituation. Still, animals on every continent are shielded from hunting in protected parks and behaviors beyond the boundaries of secluded reserves are available for contrast.

This sort of comparison was not lost on Tim Caro, who developed a project in and beyond Katavi National Park in western Tanzania. The results?

Giraffe, buffalo, roan antelope, zebra, and reedbuck were significantly more likely to flee from humans when outside parks. The variation did not characterize all species, however. Differences did not exist for elephant, eland, hartebeest, topi, impala, and duiker. In other words, some species were more flighty outside of reserves than inside, but not all.

In North America, trends are similar. Deer, elk, caribou, and pronghorn are more wary where hunted. They also adopt other tactics to thwart humans with rifles. Some individuals move more during the day, presumably to avoid areas with hunters, although perhaps they do so because they are flushed more frequently. Others become nocturnal, reluctant to leave thick vegetation in daylight. Many know the boundaries of protected regions and migrate swiftly to them.

Why Africa's big game persisted with early humans is a question that may never be resolved. But it is possible to examine how humans and predators shaped past behavior and continue to do so now. Carefully controlled experiments and natural variation in species or populations that differ in history can collectively shed light on the interaction.

WHILE CAUSES OF North America's big game collapse remain contentious, the evidence is clear that some losses cannot be explained by climatic changes alone. Colonizers, human or otherwise, cause massive ecological damage. Charles Darwin noted this as early as 1845: "We may infer . . . what havoc the introduction of any new beast . . . must cause in a country before the instincts of the indigenous inhabitants have become adapted to the stranger's craft of power."

The wave of predatory invaders comes in many forms. Brown tree snakes have decimated Guam's native bird population, just as aquatic snakes have done for the Mallorcan toad. Mongooses and weasels have taken their share of Hawaiian birds and New Zealand's flightless takahe. But perhaps the most potent virulence has been disease. During the two centuries after indigenous Americans were first contacted by Europeans, some fifty to ninety-five percent died. Such reductions had to have ecological reverberations. Human hunting pressures would have been reduced. Perhaps the bison of Atlantic forests noted from the 1700s were there only because Indian densities were so reduced. Perhaps current wolves were so widely distributed 150 years ago because dire wolves were lost, grizzly bears because of the loss of the larger short-faced ones.

Based on knowledge about how faunas have changed, a dozen conservationists and ecologists published a paper in the British journal *Nature* suggesting that the great ghosts of North America should be replaced with ecological surrogates, those species closely related in form or function to the original ones. The rationale was to enable ecosystems to function as they once did. Pleistocene re-wilding is a ridiculously simple concept. Its aim is to develop healthy ecosystems by restoring ecological interactions by returning lost species, particularly where losses may have occurred at the hands of man. No one seriously expects Holarctic lions or mammoths to reappear back on the North American prairies, but small steps are underway.

Boslon tortoises are the continent's behemoth reptilian browser, weighing in at a hundred pounds. They once lived in the grasslands and deserts of West Texas but became extinct there four thousand years ago. A few relicts survive today in northern Mexico, and now some have been returned to one of the New Mexico ranches of media mogul and conservationist Ted Turner. The condor story differs to a degree. These large soaring vultures once fed on ground sloths and Harrington's mountain goats in the Grand Canyon. That was thirteen thousand years ago. Although the two herbivores are extinct, condors are back in the skies of northern Arizona, where they consume the carcasses of surrogates: livestock, elk, and deer. An ecological service has been returned along with a species that existed long before Columbus touched a new world. Llamas, or even camels, could serve as proxies for extinct long-necked camels. Indian elephants might even fulfill a role once played by mastodons in the mesquite and acacia dominated lands of the southwestern United States.

The underlying concept of Pleistocene re-wilding has been to start small, be strategic, and understand and then recreate past ecological interactions. The scale should be modest, at least initially. As for predators of the big browsers and grazers or the speedy pronghorn, surrogates to replace Holarctic lions and American cheetahs are possible. Reality is a different issue. Lands are flush with the livestock that allow ranchers to make a living. Where wolves and grizzly bears are already unwelcome, is anyone truly in favor of seeing the return of 350-pound flesh-eating lions?

I might be, but the US sheep industry is not. Neither are some scientists who claim not only that the habitats have changed but that conservation is difficult enough without squandering funds on projects that are mere pipedreams.

The ecologists who wrote the re-wilding article think differently. They want to celebrate North America's rich ecological history with the designation of a park, a very large one. I am among the dozen authors who penned that paper. No one denies that historic niches are unfilled today. Is dreaming about bison on American prairies bad? Is combating mesquite-filled and weedy brush lands with browsers good? An optimistic vision for restoration has value. Uncovering facts and being experimental is what science is about. Why not a humble beginning—say, on a hundred fenced hectares, or a thousand, or even ten thousand? Bison re-wildling began a century ago. Bolson's tortoises have been launched. When do ghosts become real?

chapter 10

On Being Caribou and Musk Ox

We are not just protecting the walrus and polar bear and caribou, we are protecting a world, pretty much the same world that some of our reindeer-hunting ancestors inhabited. . . . A world that evolved as the Pleistocene ice sheet plowed across the continents, driving the older world of warm grasslands and broadleaf forest southward.

DAVID RAINS WALLACE (1986)

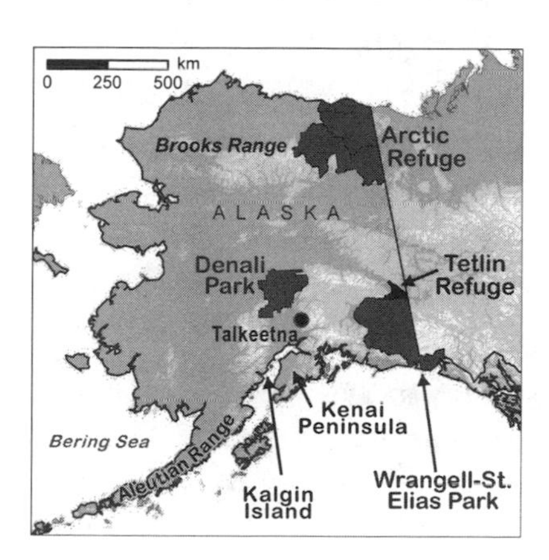

I LOOKED ACROSS a quiet expanse as lightly falling snow faded into the evening shadows. Before dawn I awoke. The sky was star studded. Sound travels further in cold and, in my semi-consciousness, I thought I heard soft shuffles passing somewhere beyond my Alaskan cabin. My watch read 4 AM. Perhaps it was a dream—images dancing across the mountains of my mind. I snuggled back into the warmth of my sleeping bag.

After first light, I went outside. The eastern mountains were washed in pink. I scanned ridgelines looking for signs of life. Square Lake did not seem the same. Slicing across its frozen surface was a swath of freshly churned snow. Nearly a half mile long, the path flared in then out, like an accordion stretching with each note.

The first caribou of the season had left their trail. By afternoon, others were trundling past. Concave hooves clicked on ice as hard as pavement. Snorts and grunts brought the frozen lake to life. Within the next few weeks, fifty thousand more would arrive. The migration would be in full swing, completing a cycle that has replenished the tundra since the melting of the Talkeetna glaciers. In this part of Alaska, I would focus on caribou responses to a major predator—wolves.

CALLED XALIBU BY Micmac Indians of the Canadian Maritimes, the name caribou means "pawer" or "shoveller." It aptly describes the feeding behavior of this gregarious animal. Xalibu obtain their primary winter food, lichens, by digging through snow. In northern Europe, the name is different; the Sami of Norway say "reino," which underscores why, even today, confusion exists between reindeer and caribou. Both names apply to the same member of the deer family—*Rangifer tarandus*—a species described in 1735 by Carolus Linnaeus: "horns branched, round . . . palmate . . . inhabits alpine mountains of America, Europe, and Asia . . . descends in winter into the plains."

Linnaeus was on target. Caribou are the world's true long-distance land migrators. Some cover more than 1,800 miles, a distance that would stretch from Mexico beyond Canada's border and dwarfs by more than three times the migration of the Serengeti's famed wildebeest.

In North America, the domesticated form is called reindeer, as it is in northern Europe and Asia. So reindeer can be either a wild species or a domesticated one, depending upon location and herd history. Wild forms are larger bodied, usually darker in color, and normally mate and give birth later than domestic ones. Adding to the confusion, both forms may roam freely. Some domestic reindeer become feral, subsisting unaided by humans. Not only do domestic and wild forms interbreed, as has occurred on all northern continents, but their coexistence carries economic and ecological consequences.

In recent history, wild reindeer in northern Russia have been harassed, trapped, and even gunned down by helicopter. There, local herders lose their domestic stock to wild bulls, and the wild deer compete for food. The same is true in Alaska, where reindeer are preferred by Iñupiat herders to their wild cousins.

Even in today's changing world, where indigenous peoples still live in remote settings but have access to modern amenities, some still rely on a domestic reindeer economy. The Sami in Finland use all-terrain vehicles and snow machines to herd their animals, while the Nenets (previously known as Samoyeds) of Russia's Yamal Peninsula rely on traditional reindeer-drawn sleds. The relationship between human culture and reindeer stretches from the Kara, Chukchi, and Okhost seas along the Pacific and Arctic oceans to northern Mongolia and up to the northern tiers of Finland and Norway.

While a herder's life of subsistence cannot be easy, there is the animal's perspective to consider. If I were a wild reindeer eking out a paltry existence from the cold, polar deserts of Greenland, I would claim that I am not the problem, arguing the opposite, in fact: that I have been persecuted and shot. I would argue that I have been chased, viciously bitten, and killed by dogs, that I had never been exposed to the painful ravages of biting flies until domestic reindeer were introduced. That is when life really changed. With the domesticated forms came the warble and nasal bot flies.

These flies have dense fur often striped in yellow, orange or black—coloration that might lead to the impression these are bumble bees. Feeders of pollen they are not. The parasitic flies deposit their eggs in the hair on the legs or bellies of caribou. Once hatched, the larvae burrow into the skin. Their breathing holes often become infected and abscessed, and the maggots emerge through holes as large as a half-inch.

Domestic reindeer have been introduced far beyond Greenland. By 1892, they reached Alaska's Seward Peninsula, courtesy of efforts from Siberia. Some escaped, and the domestics mingled with native caribou. Within forty years a peninsula that had been biologically more dead than alive swarmed with nearly half a million reindeer and caribou. They attracted wolves and grizzly bears. They dealt with indigenous human hunters, and they dealt with voracious insects.

IT WAS THE possible interaction between predators and prey in vast, large ecosystems that brought me back to Alaska. My interests lay not solely with moose. I wished to understand whether different species developed more generalized strategies for dealing with predation. I took time to look for potential study regions, some that could include caribou and musk ox.

Bounded by the five hundred-mile-long Brooks Range to the south and the Beaufort Sea to the north, only native caribou migrate to the coastal plain. Rich in land, this thirty thousand-square mile wilderness known as the Arctic National Wildlife Refuge is the summer home of red-throated loons, snowy owls, snow buntings, phalaropes, and sandpipers. It has been visited by native and non native hunters and by President Jimmy Carter. It was studied by Olaus and Mardy Murie and photographed by Subhankar Banerjee. Peter Matthiessen, David Sibley, and Terry Tempest Williams have defended this chunk of land, an ambassador of wilderness, a palace of open space.

It was mid-June when I arrived in Arctic Refuge. The bulk of the 130,000 migratory Porcupine caribou were still far to the east, in Yukon's Ivvavik and Vuntut national parks where the snows were deep. I boarded a small plane to survey the coastal plain. With me were Bill Weber and Steve Zack from WCS. Earlier we rafted the Hula Hula River, starting in the Romanzof Mountains and cutting through the Brooks Range to the Beaufort Sea. If I was going to seriously contemplate a project here, I needed to understand the biggest problem of Arctic Alaska—logistics.

Looking down, the tundra glistened—a hexagonal landscape of kettle holes and glacial fluting, cotton grass, water and snow. With 24 hours of sunlight every day, we hoped to see migratory caribou. With each tuft of air, with each turn, we bounced, looking for signs of life, signs of quadrupeds. Months earlier, a polar bear denned to our south, some thirty miles from the Beaufort Sea, out on the northern slopes of the Brooks Range. That was about when Gale Norton, then secretary of the US Department of the Interior, flew over the site, finding it nothing more than a vastness, white and barren.

Wasteland it was not. There were golden eagles and gyrfalcons, rough-legged hawks and long-tailed ducks. Buff-breasted sandpipers performed courtship displays with wings spread wide. Bluish arctic foxes hunted marmots and lemmings. Red foxes denned. A porcupine had passed, its feet and tail leaving not-so-subtle clues in the melting snow. In mud were the haunting tracks of a wolverine. Lichens, Siberian phlox, and purple anemone colored an austere landscape.

One day we heard the howling of wolves. On another, a pack crossed open tundra. There were grizzly bears and moose. Dall sheep grazed rocky slopes. We floated past braided drainages and shot rapids below sheets of ice ten feet thick. There was mammoth ivory on sand bars and combs fashioned from antlers by the earliest inhabitants of the area.

Along with Lapland longspurs and golden plovers were musk ox. With horns contoured along a thick frontal boss, this archetypical survivor of the Pleistocene watched nervously. Adeline Raboff, a Gwich'in from Arctic Village, and I nudged past the wooly beasts as strands of their fine underfur fluttered in wind.

Although tropical biomes hold more biological diversity, the mix of land and marine mammals was high for this part of the world. Ten species use Arctic Refuge's coastal waters. Regular inhabitants are bowhead and beluga whales, polar bears, and three species of seals—ringed, bearded, and spotted. More unusual visitors include walruses and harbor porpoises, gray whales and killer whales.

We watched raw beauty collide with wildness as we flew above the tundra. Sea ice and coastal fog blanketed the Beaufort. U-shaped valleys and canyons cut the Brooks Range. Glaciers sat like thrones. But caribou were nowhere to be seen.

We pushed eastward into Canada. Small pockets of animals appeared, groups of twenty, forty, and sixty. Most fed normally, but a few ran frantically. It was too early in the season for mosquitoes to have caused the panic, and it was not us from whom they fled. We circled once and banked sharply. My eyes pressed hard against plexiglass as I hoped for a better view. A grizzly bear chased a cow with a two-week-old calf. Most encounters between grizzly bear and caribou last less than a minute, but we did not want to interfere. We pulled away.

About one-third of the interactions result in a kill, and grizzlies then spend about 15 minutes eating. Federal biologists found that females with cubs had about twice the kill rate of male bears, and an estimated two to three thousand calves are preyed upon annually on the coastal plain. Modeling suggests that such rates of mortality would have a negligible effect on the overall size of the Porcupine caribou herd.

As we left the narrow coastal plain and neared mountains, more bears appeared. With less water-soaked habitat, there would be more ground squirrels and marmots, as well as greater access to plants.

In 1954, a young undergraduate from the University of Alaska wrote to Olaus Murie, offering to volunteer with the exploration. The New York Zoological Society sponsored the expedition; the youngster's name was George Schaller. From field camp, Schaller wrote: "In this land of the midnight sun the jubilant singing of juncos, myrtle warblers, tree sparrows, and gray-cheeked thrushes can be heard. . . . [F]rom our tents we see bands of caribou, . . . from fleet wolf spiders to ants, beetles, and even mosquitoes. . . . It is warm now, daytime temperatures into the sixties." Almost fifty years later, Schaller would append his initial feelings: "Arctic Refuge is a place of living grandeur, one throbbing with life, an Arctic legacy of world importance that we must treat with respect and restraint."

TALL, PREPPY, AND boyish, even at fifty, Steve Zack of WCS had once dribbled basketballs on the courts of the Portland Trailblazers. An ex-faculty member at Yale, Steve and I now planned research in the Arctic. His efforts would be on shorebirds along the Arctic's coast plain. He would succeed. My fate would differ.

"We don't have any wildlife management issues remaining. We understand what's going on here in Alaska. I think you gentlemen ought to go elsewhere, maybe Canada."

With that, Pat Valkenberg, director of research for the Alaska Department of Fish and Game's northern programs, dismissed our efforts to study what Pat scent marked as state-owned game animals. Pat's generous invitation to work elsewhere revealed an underlying anxiety—xenophobia, the fear of outsiders.

Perhaps the real problem was not xenophobia but me. I was under the impression, however, that Alaska's conservation challenges had grown more acute rather than nebulous. Global warming, petroleum development, outbreaks of bark beetles, conflicts between wildlife and humans in Anchorage, and the recent loss of indigenous woodland bison topped the list. And then there were those "damn" predators and locally declining herds of caribou and moose. I also thought that my previous and ongoing work with Alaska's university scientists and others in the local fish and game departments were demonstrable of successful collaboration. To Pat, Steve and I must have appeared delusional.

Rather than attempting an end run past the director of research, I opted to try to work with a different suite of land managers and biologists. I began with Patricia Reynolds, a scientist with the US Fish and Wildlife Service. She was refreshingly more upfront than Mr. Valkenberg.

Patricia suggested I work on caribou, Dall sheep, moose—anything but musk ox. I appealed to logic, pointing out that a project on a species like musk ox, which depends exclusively on the Arctic Coastal Plain for its very existence, could help scientists understand how energy development might affect this Pleistocene throwback and subsequently the land itself. I had thought that WCS funding could bring a lot to the science and offer a voice that federal officials could not.

Patricia was not convinced, suggesting that musk ox were her babies. They were the focus of her doctoral research, and her follow-up activities were highly respected. Despite being saddled with other commitments in her position with the US Fish and Wildlife Service, Patricia was a passionate spokesperson for musk ox. I tried one last time.

"We'd put people on the ground. We could evaluate pregnancy, observe interactions, radio collar more animals, and help to understand population trends. Wouldn't Arctic Refuge be better off if we worked together, if there were additional perspectives? What about the resource? What about the American public?" My pleas fell on deaf ears. Although Steve's project would find a home and support, the opportunity for a cooperative program between federal or state biologists and me in northeastern Alaska would not.

TWO YEARS HAD passed since the caribou migration funneled past Square Lake and the cabin in the Talkeetnas. I had returned to Denali, this time to assess caribou response to predators. Denali was an ideal site. Grizzly bears and wolves were relatively abundant and sometimes even caribou were visible.

Layne Adams, another federal scientist with years of experience, had monitored 350 newborn caribou calves for the past five years. Mortality that varied annually from thirty to seventy percent; most calves died within two weeks of birth. Only five deaths did not involve predation—two were by drowning and three by still-births. Wolves accounted for 35 percent of deaths, grizzly bears 41 percent of the total. Golden eagles were responsible for the majority of remainder.

Surplus killing by carnivores also occurred. Like the slain elk calves on Wyoming's feed grounds, dead caribou young lay strewn across the tundra. Layne once found 14 at one site. Five were battered but uneaten; three others had only a few entrails removed. This was not a one-time find. In 1982, a researcher found 34 wolf-killed caribou calves in the Northwest Territories. All were within a square mile and had been killed during the same 24-hour period. Elsewhere in Canada and Alaska, other caribou biologists have documented similar finds. Such events were not the result of a single aberrant wolf pack, since they occurred in different places and at different times. Given this sort of mortality, as well as predation on adults by wolves, I expected caribou to be at least as savvy to predators as moose mothers.

AT 5 AM, it was still dark. The March sky was emblazoned with pulses of green. It then glittered in silver before shifting amorphously into red. Below, spruce boughs bent under the weight of winter snow. A great horned owl hooted in the distance.

"Look at those strange clouds" said Kim, as she burrowed deeper into her down jacket. Trained as an economist and employed by the University of Nevada as administrative faculty, Kim was up in Alaska to help with my fieldwork. She had not before seen one of the northern world's great wonderments.

"Clouds?" I questioned. "It's Aurora borealis."

The volatile mix of solar eruptions and gas is magnified on entering Earth's atmosphere. Visible during winter in the far north, tendrils have extended as far south as Kansas and Utah. The colors of the northern lights reflected specific gases; oxygen produces red and green, nitrogen blue and violet light.

As the light dawned, we drove in search for beige butts against snowy backdrops. Just inside Denali's southeastern boundary, seven animals appeared. Contoured antlers topped the heads of four females. The early morning sun lit up their coarse, hollow hair, hair that bolstered buoyancy during river crossings. With spotting scopes, we gathered data. The "bous" ignored us, demonstrating why their Micmac name Xalibu was apt. They pawed through a foot of crusted snow to reach lichens below.

Four hours passed until another small group emerged. They ambled along the frozen shores of the Tanana River, moving between willows and spruce. Hoping for a continuous view, we strapped on snowshoes and hiked half a mile to a hill. The animals kept moving. We climbed down and began crossing the still-frozen Tanana. Abruptly, the silence ended.

A distinct cracking filled me with adrenaline. Kim had reached the halfway point, and an icy seam broke open, racing past me. I thought of my fall into the frigid Oshetna River, and jerked around to see if Kim was still standing. She was fine. It was only a big fracture, nothing serious. But the caribou were gone. The day was over. It had not been especially productive.

We returned to our temporary cabin in Denali. The habituated snowshoe hares and gray jays greeted us. One evening, we cooked pasta and drank wine. We talked of science and conservation. Gradually the conversation shifted to past relationships. Like me, Kim had a young daughter. Before turning in, we organized our supplies; the 12-volt battery to power the audio playback gear charged, and our boots and socks were dried. At 2 AM, I jolted awake. Kim was gone.

The outside thermometer registered −10°F. I assumed she went to the unheated outhouse, but after 20 minutes passed, she still had not returned. Not knowing what happened, I fell back to sleep.

Two hours later she returned, shivering. She had a stomach virus, and after heaving her guts out, just curled up on a bench in the outhouse. The shower curtain was a poor substitute for a blanket. We were both lucky that she didn't freeze to death. My behavior had been callous, and I vowed to be more responsive to people I worked with.

Before light and with no Aurora borealis for distraction, we renewed our search for caribou. By seven, we had found a group of eleven. After recording which individuals were central and peripheral, I played the sounds of ravens, wolves, running water, and red-tailed hawks. Only the raspy calls of ravens and the howls of wolves elicited strong responses.

With ravens, the caribou lifted their heads and their ears became erect. With wolves, they grouped. Where ridgelines obscured visibility, caribou moved off, often beyond our view. Over the next few days, we repeated the sound experiments on each new group. Individual responses were essentially the same—wolves and ravens caused alarm. Red-tailed hawks and running water did not. Hyenas and tigers were intermediate. If anything, the sounds of these unfamiliar species inflicted more curiosity, not fright; they rarely fled from the sounds of hyenas or tigers.

To thwart enemies, caribou rely on different tactics. They avoid deep snow, mainly because vulnerability to predators increases. They also seek open areas, since visibility for predator detection is better. Beyond these strategies, behavior may vary among populations and between individuals even within the same population. At high altitudes, pregnant females often separate from herd mates. Differing from those using the Arctic Coastal Plain, which birth in huge groups, caribou in Denali, the Talekeetnas, and other areas disperse. "Spacing out" is a word one researcher used, because predators seeking newborns must invest in much time searching for the widely dispersed animals.

Speed is also important to outdistance all but aerial predators. Unlike moose or bison, caribou rarely will fight carnivores, and then only as a last resort. John Burch, a Park Service biologist, witnessed an interaction in Denali in March, 1990:

> A large herd of caribou cows and 10-month-old calves came up over a rise and ran into the resting group of seven Stampede Pack wolves. The caribou scattered into four or five groups with one or two wolves chasing each. The caribou easily outdistanced the wolves.
>
> For some unknown reason a calf spilt off by itself and ran down a small ridge. A wolf broke off its chase . . . and began pursuing the lone calf. The calf easily outdistanced the wolf until it tried to cross a gully and floundered in deep snow . . . the wolf quickly closed the gap (now distantly followed by another wolf). The calf got through the snow and onto the next small ridge.
>
> Because of the trail the calf had made, the snow in the gully did not slow the wolf down. When both the wolf and calf were on good footing again the wolf was around 10 to 15 yards behind, the calf turned sharply right, and the wolf cut the corner and caught the calf. There was no struggle.

IN THE EARLY 1940s, Adolph Murie estimated the Denali caribou population to be about twenty thousand strong. By the late 1990s, there were but two to three thousand remaining. With these few roaming an area the size of Massachusetts during winter and with little road access, I worried about my ability to find them. Although we had conducted playbacks to numerous groups, no longer did Kim and I find new ones. It made little sense to repeat playbacks to the same animals. Sampling caribou from an entirely different region would yield more valuable information.

We headed to Tok, a small town situated just out from the northwest flank of Wrangell–St. Elias National Park and Preserve. The protected lands themselves were vast, a Pennsylvania-sized region of 13 million acres with six peaks in excess of 15,000 feet. The region's caribou herds were diverse. Not only was Alaska's only true woodland caribou, the Chisana herd, there, but so too were the more typical arctic types. Within the region were the Nelchina, the Fortymile, the Macomb, and the Mentasta herds, totaling 75,000 animals. Some were increasing in numbers, others declining. This consequent mix of population dynamics kept state biologists guessing and pundits questioning.

This was a land of controversy, an area where outsiders doled out big bucks to hunt, where indigenous hunters had unlimited access, and where state-supported wolf sterilization projects were underway. It was where trapping and shooting of most species was allowed in one form or another, and where wolves, grizzly bears, and the federal government were the ostensible enemies.

The land also held great promise. There were furbearers and upland birds, salmon and lowlands filled with crops of edible berries. It was a region of spectacular scenery and of conservation challenge. Not all lands had been locked up for private development or set aside for existence value. There were (and are) mining claims, vast fisheries, and opportunities for economic development, some of which included tourism and visiting biologists.

I was there to see if these caribou behaved like those from Denali. Just as my research on moose was carried out in various areas with and without predators, it was equally important to replicate sites for caribou. Like Denali, the Tok region had grizzly and black bears, wolves and wolverines.

In science, the true test for credibility is replication. When results are repeatable, the likelihood of truth is greater. Alternatively, if experiments—whether lab or field—cannot be repeated, results may be suspect. The uncertainty may stem from the inadequacy of samples, changed conditions, poor analyses, or a host of other factors. But when results from multiple sites are similar, the generality of conclusions is bolstered. The caribou near Tok offered opportunities to replicate responses from Denali.

A room in the new motel on Tok's east side served as base camp. I made morning coffee before first light while Kim organized data and checked field gear. Still, I was concerned about my coworker and how she would perform in east-central Alaska. It was not the heavy 12-volt battery that she was to carry, or the fieldwork that would be more demanding than Denali. Kim had run track, was an avid skier, and lifted weights. She knew about wildlife and was a stickler for detail and data. She was even fun, except when I mistakenly drenched her clothes and gear with steaming coffee. Rather, it was her torn anterior cruciate ligament, which had been surgically repaired a few months earlier. Neither she nor I knew how it would perform in the deeper snow that we expected to find in this location.

By the time we left the motel, it was already light. Clouds swirled above as the massifs of the Nutzotin Mountains dawned pink. Cloistered under a veil of spruce, frozen lakes expanded below. The temperature was near zero. We rounded a bend. Three black wolves exploded from an icy shoreline and vanished. The fleeting look was rewarding. Wolves that do not fear humans have short lives.

Those that associate danger with humans are likely to survive longer, as are their offspring. Generations of wolves from Denali, Yellowstone, and other protected areas do not flee aircraft. Those that had been harassed, collared, or shot at from above learn quickly about aerial predators. While intriguing, the continuum of behavior displayed by carnivores from habituation to fright and flight needs more attention.

Our search for caribou continued. Several miles down the road, trails of caribou tracks wound down a cliff and over steep embankments. They disappeared into the frozen abyss below. Nearby, five ravens fed on a dun-colored body. Tracks stamped in ice told the story. A speeding truck pulverized the unwary female. Polished antlers had been torn from her skull.

Among the world's species of deer, caribou are unique. Both sexes carry antlers. All males have them, and theirs are larger and heavier than the calcified outgrowths of females. For bulls, bigger antlers generally mean more mates. For females, the presence of antlers is more variable. In Alaska, more than 95 percent are antlered. The proportion in Quebec and Ontario is far less, and in one population in Newfoundland fewer than ten percent carry antlers. When present, however, females carry the bony appendages until after calves are born. Then they are shed, as are all antlers on all species of deer, male and female alike. The structures may be useful as weapons against predators. Female antlers might also serve an entirely different function. Since access to lichens is difficult during winter, it's possible that antlers promote individual dominance during feeding.

As we investigated the death scene, eleven caribou appeared. They glanced nervously at us and kept moving. We grabbed snowshoes, shouldered packs, and dropped into the valley. Kim plowed forward, trying to follow the trail of the caribou. The temperature warmed to 10°F. Anvils of sun now pierced the clouds, encasing the Nutzotin Mountains in a fiery glow. Groups of caribou marched past.

Over the next four days, the stream of animals continued, some searching for food and others in full migratory mode. We gathered data and did playbacks. We deliberately varied our positions, selecting different habitats in which to gauge responses to the sounds of familiar and novel cues.

We picked big groups and small groups, central and peripheral animals. The results mirrored those of Denali. The bous became alarmed when they heard ravens and especially wolves. The effects were stronger when animals were in the forest than when they were in the open.

Some six to ten thousand animals had passed. They traveled to higher country—tundra rather than forests. As we moved southwest from Tok, caribou of the Mentasta region appeared. This is a land of high mountains. Treeline is at 2,500 feet. Above 5,000 feet there is just snow, rock, and ice. The summits of Mt. Drum and Mt. Sanford soar to about 15,000 feet. Some caribou use sedge and lichen tundra several thousand feet above treeline, but others stay put at lower elevations, within shrub zones. Why females use habitats at drastically different elevations was uncertain. Might it be the variation in food quality, the abundance of predators, or some other set of factors relating to offspring survival and pregnancy?

Terry Bowyer and his university colleagues sought to resolve the paradox. They radio collared 79 adult females and followed their fates and those of their young. Food quality and the distribution of grizzly bears and wolves were also surveyed. In the end, they concluded that females with young chose higher altitudes, not because food was more nutritious, but because predators there were less abundant. These animals behaved more like bighorn sheep than caribou.

THE EXPEDITIONS TO Alaska had been successful. I now had solid information on caribou in two areas with their full complement of predators. But, I left with almost as many questions as answers. Would predator-free caribou differ behaviorally? Might they react like bison, ignoring the presence of predators?

I needed areas where caribou had lived without predators for thousands of years. I wanted to learn just how long they retained their antipredator behavior when they have not been prey for a long, long time. The islands of the high Arctic held clues to the answers of these questions.

chapter 11

Islands of Ice and Innocence

> The rigid polar regions slept. . . . Ages passed—deep was the silence.
>
> FRIDTJOF NANSEN (1903)

THE FAX FROM the Dansk Polarcenter indicated that work in their territory required "posting in advance a $25,000 bond for emergency rescue." This trip would differ from past ones. Greenland was remote, inhospitable, and foreboding.

Norwegian Fridtjof Nansen crossed Greenland on skis in 1888. Facing temperatures of –60°F and stiff winds, his navigational feat rivaled Henry Stanley's exploration of the Nile and Congo rivers. Twenty-five years later, Einar Mikkelsen described the polar ice cap in his book *Lost in the Arctic*: "We can see far and wide over this desolate land, so imposing in its utter lifelessness, a great white surface, hard as glass and yet plastic and ever slowly moving . . . so impressively silent."

Even today, Greenland is mostly inaccessible. With but a few coastal roads, the favored modes of transport are snowmobiles and dog sleds in all but the capital, Nuuk. Settlements have existed in southwestern Greenland for more than four thousand years, a period when mammoths still survived on Siberia's Wrangell Island. The majority of Greenlanders are Inuits, many of whom make a living hunting seal and whale. The degree of harvest remains controversial. Some argue the offtake has been unlimited, leading to the collapse of Greenland's coastal mammals.

Local economies are bolstered by tourists, among them adventurers of all kinds—hunters, climbers, and racers. There is the Arctic Circle Challenge, an overland hundred-mile, tent-catered event held each March and dubbed "The World's Toughest Ski Race." Climbers come to scale unnamed conical walls. Just as Half Dome juts above Yosemite Valley, sheer rock faces and serrated ridges rise above central Greenland's glaciers; the difference—no ancillary support, no hordes of people.

Researchers also come. Some study global warming on the ice cap, a task that is anything but simple. One effort required a US Army LC-130 Hercules transport aircraft. It flew insulated homes almost five hundred miles across the ice cap, leaving four scientists at ten thousand feet above sea level, where they worked for nine months.

The size of France and Germany together, Greenland has more caribou than its sixty thousand human inhabitants. Its largest reserve—Northeast Greenland National Park—is about 150,000 square miles, slightly smaller than Texas and Montana combined. In that area, musk ox number ten thousand, people zero.

I selected a comparatively easy study area known as Kangerlussuaq-Sisimiut. Situated inland at the upper edge of a sixty-mile-long fjord, the US military once operated from there. Intermittent flights arrive there today from Copenhagen via Gronlandsfly A / S.

In contrast to other areas, the Kangerlussuaq site had two advantages. Caribou were protected from human hunting, and wolves had been absent from anywhere between four hundred to four thousand years. Speculation was that the Melville Ice Bulge, far to the north, created an impenetrable barricade. Elsewhere on Greenland, wolf range was limited. By the 1930s, wolves were believed to have been extirpated. Although they may have recolonized Greenland from the northern coast by 1978, crossing the frozen Robeson Channel from Canada's Ellsemere Island, they never got close to Kangerlussuaq.

As I boarded the plane from Copenhagen, my mind raced. Greenland would not be without challenge. The logistics of data collection had not been solved. The trip was costing a fortune. Uncertainty about locating caribou was high. Densities were lower, group sizes smaller, and access more difficult than in Alaska.

OCEANIC ISLANDS HAVE always been unique laboratories for ecology and evolution. In 1866, Sir Joseph Hooker touted their importance:

> [They] are to the naturalist what comets and meteorites are to the astronomer, and even that pregnant doctrine of the origin and succession of life

which we owe to Darwin, and which is to us what spectrum analysis is to the physicist, has not proved sufficient to unravel the tangled phenomena they present.

Now, 140 years later, islands are cornerstones of conservation. Not only have they tested concepts about isolation and extinction, but islands themselves are critical refuges from predators. The Tasmanian devil survives into the twenty-first century in just one place, this Missouri-sized island. Predator-free offshore isles elsewhere have also kept other fauna safe, including the world's only flightless parrot—the kakapo—which persists only on such areas of New Zealand.

Although relaxed predation is a common characteristic of islands, the sword is double edged. Populations or species may be more easily sustained when they are not eaten. But novel meat eaters that reach islands gain the upper hand when prey are complacent, having not dealt with predation for a long time.

What constitutes a long time—five years? Fifty? Five thousand? For pronghorn, John Byers would argue that ten thousand years has been insufficient to produce changes in the grouping behavior that pronghorn use to thwart predators. Dan Blumstein, a professor at the University of California, Los Angeles, who studied kangaroos and wallabies in Australia, offered a similar time frame: "predator recognition may be retained following 9,500 years of relaxed selection." For ground squirrels in California's Central Valley, recognition of rattlesnakes still occurs, despite three hundred thousand years of separation.

In contrast, I found little variation in behavior among bison from the Tetons, the Badlands, Yellowstone, and Wood Buffalo, even though some populations experienced predation and others did not. For moose isolated from predation on Kalgin Island, fifty years—or 10 generations—was sufficient to delete their strong sensitivities to wolves or ravens. At this point, the image of prey responses to relaxed predation was still cloudy. In some species the retention of antipredator behavior was strong, in others weak.

Wolves and caribou have coexisted for half a million years. If there was any species whose biology had been shaped by a predator, it should be caribou. Greenland offered the opportunity to find out.

VARYING FROM ABOUT –20°F at night to 15°F above during the day, the temperatures in Kangerlussuaq were comparatively mild. The problem was wind. Generated from Greenland's immense central plateau, it had been blowing at 15 to 40 miles per hour for days. Until conditions calmed, playbacks were out.

There were other issues. Soon after stepping off the plane I slipped on the ice, smashing my Nikon. Fortunately, I had secondary support. The first was another Nikon. The second was Kevin White. Having completed his master's, Kevin now worked as a research biologist for the State of Alaska, and had rearranged his schedule to help.

In this village of 450 people, services were understandably limited. Though I had tried to arrange a local recon flight to survey for caribou, the pilot was in Denmark. From the States, I had also reserved a four-wheel-drive truck. When we arrived, however, there was none. We managed instead to salvage an old VW rabbit. If Kevin and I wanted to see even a snippet of Greenland, it was this or nothing. We piled an enormous volume of camping, survival, and technical gear into the tiny backseat.

The edge of the world was now ours. Beyond Kangerlussuaq, a full twenty miles of ice-bound road awaited exploration in our Hummer-proxy. Our first destination was the highest point above a frozen fjord. We hiked the blustery slopes that rose several thousand feet. Through a lattice of ice and snow, rock and open ground, we pushed higher. Coarse grasses and dwarfed willows fluttered in the breeze. Only the canyons and deep gullies held much snow. Ditching snowshoes, we chose the exposed open areas for our ascent.

On the fourth ridgeline, cloven tracks typical of a big deer appeared. At least one caribou existed in Greenland. Moving through a boulder-strewn arena, the air calmed. A guttural "pllllough" shattered the silence, jolting us from our climbing trance. Instantly, I flinched.

Four musk ox bulls surrounded us, the nearest less than ten yards away. Lethal upturned horns were lowered beneath powerful necks. Skirts of long brown hair hung almost to the ground. My focus was on their eyes, eyes gazing intently on us. Slowly we backed away. From a rocky outcrop, we took a more relaxed view.

Massive bodies appeared in our spotting scopes. Thick white stockings disappeared under the long stringy outer fur of the musk ox. Weathered, potent horns tapered to finely honed points and curved outward. Above their eyes, horns thickened to a massive boss, perhaps eight inches thick. Although both sexes have horns, those of females are thinner and smaller, perhaps deadly. Adventurer Knud Rasmussen reported an unanticipated use of horn in the 1920s. A man with an artificial leg made partially from wood and caribou skin had his foot covered by a musk ox horn.

The shaggy animals have but few enemies. Beyond humans and climate change, only a single serious one remains—wolves. Natural selection has fashioned two related strategies. First, the oxen form groups. Where wolves are more abundant, group sizes increase, probably because more ears, eyes, and noses enable better detection of enemies, or at least because the risk of predation is diluted. Second, oxen form a circular defensive array with calves often sheltered in the center. This renders penetration of the inner circle more risky for wolves. It is when the oxen flee from their enemies that their susceptibility increases.

A 1999 observation by biologists Dave Mech and Layne Adams underscores their vulnerability:

> Two wolves ran at three musk ox that fled in a tight group 90 yards upslope. A wolf bit and held onto the rump of a fleeing cow that wheeled around. The wolf's grip was lost. Both wolves then attacked the neck and face, sometimes hanging on to the nose, much like wolves have done of moose and African wild dogs of zebra. As this occurred the other wolf bit at other body parts. After about five minutes of sustained attack, the cow succumbed.

The habit of forming circular defenses has nevertheless worked well over evolutionary time. Resistance is one thing, armed humans another. A 1904 account reflects the brutal reality:

> Divided into two parties we went down each side of the flock. The calf walking a little above the others together with a cow it had sucked a little before . . . sprang up to the others, that instantly formed a circle around it. . . . All the animals turned their heads outward. . . . Snuffling and snorting, their noses turned toward the earth, the animals were ready to receive the enemy. . . . By the time the first two animals fell, it was strange to see a trembling go through the herd, the animals pressed closer to the calf, by the next 4 shots, 3 animals fell and a bull went away shot in the lungs. . . . The two cows now left pressed the calf between them; then one fell bleeding, the other protected the calf. . . . We now tried to chase the left cow on to the beach; she had sought for cover beside the killed bull and the little calf pressed closely to her. The cow walked a little but then turned toward us, fell and died. The calf took up its position at the hind legs of the cow . . . but it slipped away again [at our approach], still it ran only around the cow and it was not long before it was tied.

In addition to wolves and humans, musk ox contend with grizzly bears. In northern Alaska, both calves and adults have fallen prey.

THE MUSK OX that Kevin and I watched soon returned to feeding. Seventy yards away were caribou. The two species ignored each other, all foraging nonchalantly. After an hour we were cold and warmed ourselves by hiking uphill. Suddenly, our jaws dropped.

We stood in silence pitted in an unforgettable landscape—dwarfs against grandeur. Polar ice spread to the horizon. Raw, pure and immense, the continental ice cap generated its own wind, its own weather. Rising above the surface were nunataks—islands of rock poking above the ocean of land ice. There were ramparts and crevasses, some jade green, others jet blue. Nothing in life had prepared me for this.

Peering into the exquisite void, I remembered people who shaped my life—my parents, friends, and past mentors—all who had instilled within me an appreciation for the wonderment of our planet. This was rawness unfolded beyond my wildest imagination. I wondered what it might mean to my trusted African friend and past tracker, Archie Gawuseb.

Growing up Damara on the gravel deserts of Namibia, Archie had had few advantages. He lived on dirt floors in mud huts. He had not finished school. He had never seen a river and remained skeptical of reports of flowing water. He saw his first movie with me, and he once asked if the earth was really round. I remembered Archie's toothy grin when viewing his first river along the Angolan border. I remembered his airplane trip to America and his subsequent laughter and sparkling eyes when, at 37 years of age, he first touched snow and tossed his first snowball. These were not things he anticipated or could have imagined.

From atop that unnamed ridge, I tried to understand what Archie must have felt. Whether poor or rich, privileged or not, there were things that defied imagination. And here I was, looking at one.

DAYS PASSED, AND Kevin and I managed to find a few caribou, but the wind was still too strong for playbacks. On this brutal day, we had already hiked five and a half hours before glimpsing a solitary reclining caribou. We measured the 427 yards to it with a laser rangefinder. With no antlers, it was likely to be a female, but because males may also be antler-less at this time of the year, we were unsure. The wind had momentarily stopped. I was in serious need of data, and we opted to wait. An hour ticked by, then two. Finally, the animal stood, then squatted to pee. It was a female. We readied ourselves for playbacks.

Against a backdrop of gray rock and dreary lichens, sound after sound was directed at the feeding animal. She was totally insouciant. *Was she comatose?* Perhaps she just did not hear the sounds. When the howler monkeys hit, she finally looked in our direction. This was good. Had she disregarded all, it was possible that none of the frequencies had reached her. With a single datum under our belts, Kevin and I smiled. We hiked the tortured karsts below the ice cap, hoping for more animals.

I carried the speaker, playback unit, and amplifier; also a down jacket and vest, and an outer shell to break the wind. In addition to the gloves, I had mittens and liners, as well as another shell for pants. I also packed a Petzl lamp for emergencies, in case we remained out after dark. A book on Arctic exploration was for downtimes. I also carried two hats, a face mask, gaiters, and a tarp. There was also sun protection, a spotting scope, a tripod, and camera equipment. Kevin carried a similar load, but his also included a 12-volt battery, a spool of speaker wire, the rangefinder, and two emergency kits, one for humans, the other for equipment.

Some days were good, others not. Eventually our routine solidified. We usually left early in the morning. If it was not excessively windy, we would remain out until nine or ten at night. Back at camp, we would devour dinner and collapse into sleeping bags. After a few days of total exhaustion we would sleep in and then begin the routine all over.

Our success increased. On two very good back-to-back days, we encountered eighteen different caribou groups. We did playbacks on all, the distances from us varying from 90 yards to 810. Sometimes, as we hiked, dark shapes appeared from behind house-sized rocks. Befuddled, we just shook our heads. I would have to pinch myself to make sure that the musk ox were not illusions.

One day as we rested atop a pinnacle, I looked at Kevin. My good-natured comrade writhed in pain. His upper premolar had cracked. Options for dentistry were limited. We could try to find a flight to the faraway town of Nuuk or return to Copenhagen. I suggested we leave. Kevin thought otherwise. Committed to collecting data, he was bent on testing his threshold for pain.

As we strategized, four tiny shapes appeared across the frozen fjord. It was a human family, two adults and two small children. They constructed a miniature shelter and cut holes in the ice. They fished, returning occasionally to their refuge for what we suspected was warmth. Other families arrived, some by dog sled.

To our west were 17 musk ox and two caribou groups. This was our fourth mixed-species group. Assemblies of feeding ungulates aren't uncommon nor are different flocks of birds. Whether such foraging associations are random, adaptive, or mere aggregations concentrated on available food is uncertain. I assumed the latter, since our limited sample always occurred in snow-free zones.

It was getting late. The sun congealed into a huge orange ball and slipped below a crystallized horizon. The ice cap faded from pink to azure, and finally drifted into a deep blue. Below, two arctic foxes barked. A raven pair nested on a cliff. Our sample of vertebrate diversity had grown to seven. Caribou and musk ox, arctic hare and fox, ptarmigan and raven were regulars. Snow buntings had recently arrived.

We had accumulated more than 175 data points. For dinner we celebrated. I surprised Kevin by unwrapping yellow corn tortillas and with a can of diced green chilies. Earlier I purchased red and green peppers from the *bayukult* (grocery store), and I had brought salsa from the States. With sliced cheese melted over black beans and rice, we enjoyed burritos with a bottle of red wine. After six servings, I was full. Kevin needed nine.

The next morning we repaired gear and tidied up our data. Before noon, three Inuit mushers dressed in polar bear pants and with caribou hides strapped to sledges showed up. One spoke a little English. They were surprised to see Americans. Their sled dogs eyed us nervously. The only caribou they had seen recently were bulls. Kevin and I would search for females in the opposite direction.

As our sample sizes grew over the next few days, we talked of an early departure. Air travel to Copenhagen was unavailable for less than $5,000, and rather than spend more grant dollars, we decided to wait another week for our scheduled flight. One night, a C-130 transport plane was on the runway. The words "US Army" were stamped on its sides.

I rushed to the local hangout. An American computer specialist and logistics assistant named Tom Quinn, with whom I had once corresponded, sat quietly enjoying a beer. I learned that the plane was leaving at 6 AM.

"How do we get on it?" I asked more brashly than I really intended.

"Good luck" Tom offered with a wry smile.

"No, I'm serious. Our project is government sponsored."

"Chances are 99.9% against you. Don't count on it. Plan on Denmark next week instead. I'll ask."

Around midnight Tom had an answer. "This is your lucky day; be there by first light."

Before turning in, Kevin and I discussed our findings. The Greenland caribou had not been wary like those in Alaska. They did not respond to wolves. They did not respond to ravens or hyenas or running water. Howler monkeys drew some attention— perhaps curiosity. Certainly it was not fear, for they never ran. Group sizes were quite small, slightly more than five. Almost thirty percent of our observations had been of solitary females or couples, something we had never seen in Alaska. Eric Post, an ungulate climate specialist from Pennsylvania, had once studied caribou in Greenland and western Alaska. Although he did not conduct playbacks, his findings pointed to differences in behavior as a consequence of a loss in predation. Alaskan caribou were about six to ten times more attuned to the possibility of lurking predators. Our results were encouraging, even if they came from a single site—Kangerluusuaq. I would look further north for an additional site to see if the findings could be replicated.

At 5:30 AM, we boarded the C-130. Our trunks, playback gear, batteries, duffels, and packs were neatly strapped in, among and behind servicemen. Once airborne, the commanding officer walked back to check on the only civilians—Kevin and me. Before finishing my thank you, she interrupted, "we're all on the same team," and walked away. We touched down at a military base north of Albany, New York.

THE HEADQUARTERS OF WCS are in the Bronx. Bill Weber, the director of WCS's North American Program had suggested years earlier that if I wanted a position with them, I would need to base myself in New York. I was not sure I was ready to leave tenure in Nevada.

Another year at the university had passed. When the semester ended, I shifted to Wyoming, preparing for a new field season. With me was Kim. Our friendship had blossomed, and we moved in together. She had resigned her position at the university. So had I.

Friends and colleagues thought we were foolish to leave academic life, job security, and tenure, especially for my new position with the WCS. I was not so sure. Rare were the opportunities to combine science with conservation.

With Emmy-winning talent and global operations, WCS had done more in the last century for on-the-ground conservation than most. My new colleagues—Bill Weber, Amy Vedder, Alan Rabinowitz, and Steve Zack—had battled in trenches from Rwanda and the Republic of Congo to Myanmar and Madagascar. There were the established, like Josh Ginsberg and Kent Redford, and those hitting full stride—Eric Sanderson, John Goodrich, Justina Ray, and Jon Beckmann. George Schaller, John Robinson, Billy Karesh, and Steve Sanderson were the anchors.

Jackson Hole would be Kim's and my new base as WCS expanded its North American program. There were two bonuses. The first was personal: Sonja. She and Carol had moved to Moose, and Sonja could now live with us for part of the year. The second was professional: the Yellowstone region was to become a focus for WCS.

Some of that could wait a few months more. It was late February. Cold-weather gear and field supplies filled our living room. Soon Kim and I would head far to the north. We hoped to answer the question: does the behavior of wild reindeer on a predator-free arctic archipelago mirror that of Greenland? If responses were dissimilar, predation might be less of a force in caribou society than was generally believed. Alternatively, if reactions were similar, it would be "easier" to argue that the roots for behavioral differences, and possibly cultural variation, lay in understanding the overt and subtle effects of predation. The operative word for us was "easy."

SITUATED BETWEEN THE Barents and Greenland seas and about one hundred miles from the North Pole, Svalbard typifies polar regions. Winters are dark and cruel; ice covers much of the land. Meaning "cold coast," Svalbard is rugged and wild. Glaciers cover sixty percent of it. With four major islands and a suite of small ones, the total size of Svalbard is similar to that of the Greater Yellowstone Ecosystem. Only three mammals live on Svalbard's icy surface—wild reindeer, voles, and arctic foxes.

The vanguard are wild reindeer, numbering more than ten thousand. Diminutive in size and with stunted legs and stocky bodies, they are the lightest of the world's subspecies of caribou—only about 110 pounds, runty, yet powerful like bulldogs. Coats are light brown, bellies striped. The outer fur brightens to a pale gray or yellowish hue in winter. Like many populations restricted to life on islands, body size is reduced over time. The extinct miniature people of Flores Island in the South Pacific are one case. The mammoths of Wrangell Island another. Svalbard reindeer are no exception, a beautiful example of dwarfism.

These squat nonmigratory deer survive the cold by adding gobs of fat. Like musk ox, Svalbard reindeer are loathe to run far. Here, long distance locomotion coupled with speed is not a good physiological mix. These little fatties may have little ability to deal with the production of excess heat. Lacking in predators for somewhere between five to forty thousand years, there has been little incentive to flee, much more to grow chubby.

Svalbard's only town is the hub of Longyearbyen. With 1,500 residents, polar bears outnumber people two to one. The population of *eisbjorn* (ice bear, in Norwegian) is about five times greater than that of grizzly bears in the Yellowstone system. And they are fatally dangerous.

In the vastness of Canada and Russia, ice bears have killed 26 people in the last three decades. By contrast, in the same time span, bears killed five people on Svalbard. Four had bites to the neck and head. Predation? If there was a place where people needed to adopt the behavior of prey, Svalbard might be it.

Polar bears are consummate meat eaters. As in Alaska, the *eisbjorn* from Svalbard consume ringed, bearded, and harp seals, even walrus. Polar bears also cross open ground comfortably, which is why human Svalbardians fear *eisbjorn* and why loaded weapons are legal and common even on the outskirts of Longyearbyen.

Since Svalbard bears eat people, it was confusing to think reindeer would be immune. The best evidence for little or no effects emanates from long-term studies. In 2003, Jos Miller and colleagues reported: "Svalbard is free from predators, so survival is largely dependent on food availability and weather."

Polar bears, however, are not automatons waiting idly for seals. In Greenland, they have taken down musk ox and have been seen chasing caribou in Canada. On Svalbard, Norwegian researchers inferred predation on reindeer:

> In late March 1995, in the pass between Ringdalen and Tufsdalen at 500–600 m above sea level and about 20 km from the coast, tracks were observed of a polar bear traveling northward through 40–50 cm of fresh snow. . . . The tracks lead directly to a dead reindeer carcass that was still warm and about half consumed.

Thirteen cases of scavenging or predation were reported, a listing that would cause any breathing human to think about safety. Whether such events seriously compromised reindeer behavior and responses to predator cues is something that I was determined to discover.

MORE THAN THREE hundred pounds of gear in trunks labeled "Norsk Polarinstitutt, Longyearbyen, Norway," were ready to go. It was March 18, 2001, and Kim and I would soon head to the high Arctic. My carryon luggage for the plane would include binoculars and cameras, cassettes of animal sounds, film and data books, a GPS unit, and extra batteries. We worried whether our half dozen trunks and duffel bags would survive the five flights needed to reach Svalbard.

After 36 hours passed, only two flights remained. In the airport at Oslo, we pushed a cart filled with our heavy payload. I fumbled for paperwork as a nice-looking customs officer questioned us. He was more interested in reading my eyes than in my documents.

"What will you do in Svalbard? What is in the trunks?"

"Gear for playing the sounds of gulls and other birds to reindeer."

The man looked at me, and then turned his attention to Kim, slowly moving his eyes up and down. Satisfied that the bedraggled couple before him was for real he said, "Its' cold up there, have a good trip."

Shafts of light illuminated snowy mountains as we descended into Norway's most northern city, Tromso. The island of Svalbard lay only a few hours north. Deep, icy seas turned to rugged peaks poking above surfaces polished in ice and snow. Forty-eight hours had passed since leaving Wyoming.

We deplaned into Longyearbyen's warm air terminal. All the trunks had made it, as had our cold-weather and survival gear. Les than a mile away, two reindeer grazed below rugged slopes.

As a door swung open, we stepped from the airport into an outdoor freezer. With wind chill, it was –30°F. We pulled on neoprene masks. Our next concerns were finding a room for the night, securing maps, meeting with government officials and officers, locating snow machines for backcountry travel. We walked a mile and a half to Longyearbyen. Inside its four mountaineering stores were reindeer skins, seal hides, gloves from arctic hare, even clothing made from polar bear. The most common items were familiar brands—Patagonia, Mountain Hardware, North Face, and Arc'Teryx. Outside, it was still light, gaining at a rate of twenty minutes a day. From our room, we peered up a glaciated valley hoping for more groups of reindeer.

"DO YOU KNOW how to use this 307?" asked Reinar, handing me the butt of the rifle and shoving cartridges into my hand. "This is for *eisbjorn,* which you must take very seriously." Next came an emergency locator beacon and a two-way radio.

More supplies were piled next to our trunks. There were red flares and a pistol for shooting them. The yellow cartridges were cracker shells, designed for scare tactics. We needed to be certain to shoot in front of the bear since a warning shot overhead was more likely to trigger a charge.

Reinar proceeded with the training session. The snow machines were not the best, but they were all that was available. As if this scenario was not already James Bond–like enough, Reinar proceeded. After another five-minute demonstration, we were deemed expert mechanics—starting the machines at –40°F, navigating slopes with a heavy sledge, and adding oil and changing belts. I thought the lesson was over until Reinar added a last thought.

"Remember to carry the gun with you at *all* times." He repeated the word "all," and then continued. "Unless you go into the bank; then please remove the gun and leave it outside."

Had this not been enough to get our attention, his final comment did. "It's getting late. You have many miles to go. You need to get to the cabin before it is dark. It is going to be cold. You don't want to freeze to death."

Kim and I looked at each other, incredulous. Even if we dared recount this scene back in the States, no one would believe us. A government official hands us two guns, munitions, a cavalcade of fireworks, and tells us to carry it everywhere except the bank. Find the cabin—actually a point on a map—before it gets too cold, otherwise we would die painfully. From here, this made total sense. Polar bears were dangerous and occurred miles inland, even on mountaintops. Prudence was, we supposed, the better part of valor.

Our study area was called Nordenskjöld, and the cabin was just 35 to 40 miles distant. Kim plugged coordinates into her GPS and shoved the unit into her chest pocket for warmth. At –10°F, batteries did not work well. But, if Nansen could navigate oceans without a GPS, surely we could find a cabin. We could either cross a steep glacier to find it, or go the long way around fluvial gravel plains and moraines.

We opted for the latter, knowing that only a few days earlier a man had died when his snow machine plummeted into a crevasse. Since both Kim's and my machines would tow sledges with an additional eighty liters of petrol and our mass of field supplies, neither of us was anxious to test our adeptness while pulling 350-pound loads over precipitous inclines.

By midday we were off. The comment "that the stove in cabin can be difficult" still nagged. If it did not work, the challenges would grow more interesting. Svalbard had some vegetation, little of it woody and virtually none taller than my ankle. With nothing to burn, there would be no warmth. At least I had my trusty multifuel mini-stove. Like a faithful friend, it had served me well in Russia and Greenland, and it would always convert ice to a warm drink.

An hour before dark, we entered a valley surrounded by gentle ridges and impressive mountains. The cabin was on a snow-covered bench. To its southwest was Van Mijenfjorden, a spectacular ice-filled fjord that cut Svalbard's largest island, Spitsbergen, almost in half. Close to the Gulf Stream, we were in Svalbard's banana belt.

Inside the thickly walled cabin, the temperature was –2°, ten degrees warmer than outside. With cold fingers from the ride, we tried to light the cabin's stove. The paraffin was not cooperating. I pumped up my mini-stove and lit a match. A gasket split, sputtering fuel everywhere. Its usefulness was over.

We both moved closer to the paraffin stove, wishing the inert metal would warm us. I popped the top. Paraffin doused my face, burning my eyes. Kim finally succeeded in lighting the bloody stove. Two and a half hours later, enough snow had melted for a lukewarm dinner. By the time we climbed into sleeping bags, the cabin was a toasty 16°F.

Thirteen hours later, we awoke from deep slumber. It was sunny, the outside temperature registering at –17°F. The paraffin stove was again difficult, requiring another two and a half hours just to heat water. We expected Svalbard to be miserable, and I couldn't help but wonder why Kim had been anxious to come. Perhaps she felt left out from the mission to Greenland and secretly desired to experience the Arctic for herself. Or perhaps she just wanted to help.

We prepared the snow machines, filling them with oil, and maintaining optimism that if we found reindeer the battery would have enough juice to power the playback unit. By the time we were ready to leave, the sky had collapsed. Clouds obscured peaks and descended rapidly into the valley. We left anyway. The light deteriorated. Cliffs merged with crystallized fog. The sky glued itself to the ground. With visibility only a few meters and the unexpected thrill of hidden precipices, we limped the last five miles back to the frigid cabin.

Riddled with pain from the cold, her fingers and toes white and numb, Kim climbed into her sleeping bag. I finally managed to eke out a few BTUs from the paraffin stove and then climbed in with Kim. She pressed her bare feet onto my chest, and fingers into my back. The next morning, we would have to again try for data.

Before starting the machines, we checked oil levels, only to discover that they were filled to twice the recommended amount. Somehow, we had overfilled them the day before. At –20°F, we began to extricate the excess sludge-like oil. Using pipettes, we pinched a droplet at a time. Numb and hurting, I took a break as Kim looked for a different solution. When I returned she was sucking more oil in a single breath than I had in ten minutes. Her secret was large plastic tubing, fashioned into a straw; we removed the excess oil in an hour.

The skies were clear. With sun on distant mountains, we skimmed across snow in long shadows, the temperature at –8°. Five reindeer appeared, three were adult females. We readied for playbacks, first recording snow depths, topography, and the distance between the animals and us—160 yards.

The pint-size animals fed intently, ignoring the sounds of water and wolves. They became vigilant when I played the calls of Golden plovers and Glaucous gulls. With only a single sample, I didn't want to make too much of the reactions. Perhaps the reindeer had simply wanted to welcome the sounds of returning avian migrants.

My selection of plovers and gulls for playbacks rather than ravens and red-tailed hawks was a simple fact of geography. Neither ravens nor red-tails had ever made it to Svalbard. The scavenging role of ravens had been replaced by Glaucous gulls. Plovers, which do not associate with carrion, offered a control.

The next day the wind howled. Inside the cabin we had perfected the stove and the temperature reached a pleasant 42°. Kim read Aspley Cherry-Garrad's *The Worst Journey in the World*, an account of a group that splintered from Scott's 1910 expedition to the South Pole. Cherry-Garrad put up with temperatures as low as –70°F, reindeer hides and clothing that froze so solid—some from their own sweat—that it took hours to peel apart, and navigation of crevasses during the starry polar winter. When all was said and done, even the emperor penguin eggs that Cherry-Garrad almost died to obtain were unwanted when he delivered them to British Museum. By contrast, our situation was cushy.

Finally, the wind calmed. We fired up our machines and searched for reindeer in an area called Semmeldalen. We found none. For lunch, we ate a mix of peanuts, pumpkin seeds, and power bars while sitting on our sleds. The water in our Nalgene bottles had already frozen, but the lack of fluid meant less need to remove four layers of outer clothes to pee.

We rode up a side canyon and entered a large valley. Five groups of reindeer were spread over several miles. We broadcast an array of sounds. Unlike the earlier playbacks, these reindeer ignored all, gulls and plovers included. A few looked toward the speaker, but none broke off their foraging for more than a few seconds. Howler monkeys elicited no more of a response than wolves.

In between our efforts to gather data, we attended to other issues. The spates of snow blindness that I had been experiencing were intensifying. Sometimes, all I saw was a foggy haze. My goggles, which had serious leaks, offered little help. Air blasted my unprotected skin at the equivalent of –30°F. Although my hooded neoprene face mask worked, my exposed facial tissues were becoming raw. Areas above my eyebrows and a small portion of my forehead were peeling off. Reindeer had natural resistance; I did not. The early stages of frostbite set in.

One evening, the wind gusted at thirty to forty miles per hour. Outside, the ambient temperature dropped to –25°F. We needed water, no longer an easy chore, as clods of dirt were mixed into the snow. The good stuff required scraping the surface clean and separating dirt from reindeer feces. As if prepping for a fancy banquet, Kim dressed for the occasion. She pulled on a snug sweater, and yanked up overalls, then a heavy coat and parka. Glove liners and mittens, a hat and head lamp were next. Without a face mask she closed the door. After seconds outdoors, she grimaced in astounding pain, her face experiencing the equivalent of –62°. But at least we would have morning coffee. As the intoxicating smell of rich grounds wafted throughout the cabin, we reminded ourselves just how close to nature we were. The dark, bobbing chunks in our water were reindeer poop.

OVER THE NEXT few days, we expanded our travels, finding more and more reindeer and conducting playbacks to animals on slopes, near cliffs, and on windswept plateaus. Animals ignored the sounds and pawed at the deeply crusted snow, digging craters for lichens and other food. Some spent only a few minutes in excavation, but others committed hours.

The costs of acquiring food by what has been termed "cratering" proved substantive. Using tame animals, Alaskan researchers placed heart rate monitors on caribou and subjected them to food buried under different snow conditions. Where snow crust was hard but only an inch thick, animals broke through with a few leg swipes. When the surface was more hard packed, they could not smell lichens and searched further and wider. The cost of digging per leg stroke varied four-fold, from 118 joules in soft snow to 481 when it was firmly crusted.

Snow craters were of clear importance to reindeer and offered clues about the role of dominance in food acquisition. During winter, when most males have cast their antlers, female rank typically improves. Studies in Finland revealed antlered cows enjoy enhanced access to craters for themselves, their calves, and their gestating fetuses. Calves sharing craters with their antlered mothers had less weight loss than those who did not.

By packing the surface, our snow machines could easily increase the energetic costs to the reindeer we studied. To minimize additional stresses, we always attempted to use the same over-snow travel routes. This rendered not only a smaller trail system but one with greater travel efficiency, since we no longer plowed daily through pockets of deep soft snow.

APRIL 2, 4:40 AM Daylight lasted for 24 hours. We planned to go to Akseloya, a remote island at the western edge of frozen Van Mijenfjorden. Ringed and bearded seals were there. So were polar bears. Arctic foxes made a good living from their kills.

We were to pass a dozen glaciers and cross the largest, a 12-mile-long swath, through a delicate pass two thousand feet above the frozen ocean. I slept poorly, remembering the fatality in a crevasse earlier during our visit. The haunted writings of Valerian Albanov did not help. In 1912, this 32-year-old Russian explorer lost his ship in the ice-covered Kara Sea. In his book, *In the Land of White Death*, Albanov wrote: "The surface of the glacier was as smooth as a mirror and covered with snow, permitting the sledge to glide easily across it. I had drawn the men's attention to the dangerous crevasses that could be hidden beneath the snow."

Kim and I reached the glacier at six in the morning. The ambient temperature was –34°F. For a brief second, she glanced my way and then powered forward, racing higher and higher in what seemed a near-vertical rise. She disappeared through the notch-like pass. I was happy she had gone first.

Halfway to Akseloya, we had already counted twenty reindeer groups, none with more than five individuals. The swollen paw prints of *eisbjorn* soon appeared, some virtually atop those of reindeer. *Did the bears track reindeer or was the converse true? Was it a mere coincidence that the two species occurred together or was this simply a good travel route?* My senses heightened. I checked the chamber of the rifle strapped to my back. Kim touched her holstered flare pistol, assuring herself of our second line of defense. The first were our eyes. I thought of the power and stealth of the only polar bear I had seen in the wild, just off the pack ice on an island in Alaska's Beaufort Sea.

My goggles still fit improperly. Frigid air attacked my unprotected skin. Annoyed into irrationality, I replaced the goggles with sunglasses, hoping they might help. Twenty seconds later, the goggles were back on. Though imperfect, more of my face was shielded. Half a dozen skuas appeared, the first we had seen.

Along the fjord, ice ridges and pressure floes were stacked like car wrecks, jammed against the rocky shore. Some were the size of small buildings, others mere lumps the shape of ice bears. This was a new micro-habitat—forests of ice. Chunks were scoured blue, blue-green, and green. One hundred years earlier, Fridtjof Nansen described ice floes.

> Making one's way through these new ridges is desperate work. One cannot use snowshoes, as there is too little snow between the piled up blocks of ice, and one must wade along without them. It is also impossible to see anything in this thick weather—everything is white—irregularities and holes; and the spaces between the blocks are covered with a thin, deceptive layer of snow, which lets one crash through cracks and pitfalls, so that one is lucky to get off without a broken leg.

Polar bears never appeared. Tracks meandered through stacked ice and over fissures until covered by the wind-blown snow. The ambient temperature had risen to –15°F, the wind chill bringing it down to about –40°. Later, as we enjoyed the heat of the stove back at the cabin, the lowering sun turned the landscape a brilliant pink. A solitary bull disappeared over a ridge.

RONNY AANES INVITED us to dinner. Two years earlier, we had met at a conference in Trondheim, where he first planted the idea of my research in Nordenskjöld. Ronny was completing his doctorate on Svalbard reindeer. We now convened at a hotel in Longyearbyen. He ordered *minke* (whale) and ringed seal. I followed his lead. Kim avoided the mammal dishes, choosing arctic char, a trout relative.

Ronny peppered us with questions, anxious to learn what our playbacks taught us about reindeer responses to the predatory void. It was simple. Few animals reacted to the sounds, whether wolf, Glaucous gull, Golden plover, or howler monkey. They did not engage in false alarms, nor did they group up. Although membership in groups is viewed as a way to enhance predator detection, 18 percent of the reindeer on Svalbard and 30 percent in Greenland were either solitary or in couples. By contrast, Alaskan caribou were always in groups of three or more. The lack of reaction by Svalbard reindeer to playbacks could not have been due to an incessant need to eat during harsh and cold winters. Regardless of global warming, caribou from Greenland and Alaska did not exactly enjoy winters of luxury. The responses of animals from Svalbard mirrored those from Greenland, the lack of predators being the most likely explanation.

I had taken the topic of relaxed predation about as far as I could. Other questions persisted, many related to learning, some to culture. These were not easily addressed on caribou. Rather than contrasting populations, as I had done among Greenland, Svalbard, and Alaskan sites, I wondered about the type of variability that existed naturally within populations. Would females who lost calves alter their birthing sites the following year? If they did, would it be because they lost their young to predators, per se, or might it merely be that the loss of a neonate to any cause triggers movement to different birthing locations? Did mothers teach their young?

Answers would shed light on how behavior is modified by experience and other events. My recent work had concentrated on caribou, but it had not been fine-tuned. I did not know the histories or personalities of individual caribou in the Arctic or in Alaska.

Elsewhere I did. The longitudinal profiles of moose in the Tetons and the Talkeetnas held great promise. I would be able to derive insights from contrasts within and between populations.

part iv

The Predator's Gaze

[N]earby a snow leopard glides among the crags, my mind sometimes drifts back to thoughts of the Pleistocene and its extinct carnivores. I wonder how the lithe hunting hyenas, the lion-sized giant cheetah, the cave bear, the spectacular array of smilodont or sabertoothed cats, and many others, all now vanished, once fit.

GEORGE SCHALLER (1996)

chapter 12

Changing the Rules of Engagement

When a predator enters a landscape and encounters prey that have no previous experience of that predator, the prey can suffer heavily. . . . Most of the damage is inflicted along the front of the advancing predator population.

AMERICAN ASSOCIATION FOR ADVANCEMENT OF SCIENCES (2001)

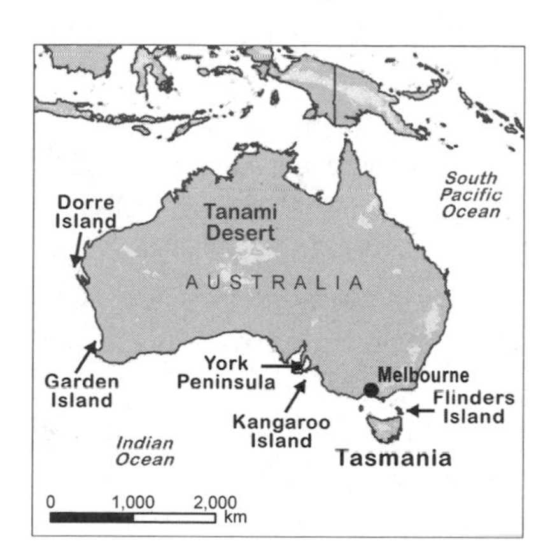

PREY AND PREDATORS share special relationships. Some remain relatively stable; others change rapidly. The gaze of a snow leopard may give a mountain-living goat reason to pause, but when a new predator with new hunting skills arrives on the scene, many species have little or no experience in surviving it.

There are different ways to explore prey and predator relationships. In a few instances, prey have expanded into areas that lack effective carnivores. Then, we can ask whether prey differ in responses to recently extirpated or long-extinct predators. Other sorts of insights can be gleaned when wildly exotic species invade. Sometimes they are introduced animals known as aliens. They can also be humans armed with new technologies. With these issues in mind, I spent time in Africa, Australia, Argentina, and central Asia; my goal was to appraise how prey and predator relationships have changed.

THE SWIFT PACE of environmental alteration creates overwhelming challenges for many large animals. Options for adjustment are few, notably because the length of a generation for big beasts is so long that specialized adaptations are rarely possible. Just like with climate change, many species may simply not be able to change in time. Additionally, the spectrum of forces driving extinction—habitat destruction, alien species, overexploitation, and secondary extinction—limits further chances. Some scientists have boldly suggested that behavioral naiveté be added to the list.

Also known as "ignorance of danger" or "innocence," the state of prey unpreparedness is something that big game hunters in the American West continue to fret about. The concern has focused more on reintroduced predators rather than on human technologies that include more lethal weapons and increased access to game via all-terrain vehicles. The issue is strikingly simple. It does not concern whether recolonizing carnivores will drive prey to extinction; rather, it is about whether carnivores and humans compete for the very same prey. Nonetheless, there is a different point that has been omitted from the broader debate—the rich biological history of predator-prey interactions. How do prey species deal with stability and change, particularly when familiar predators are replaced by novel ones, and when hunting methods differ?

Black rhinos offer a beautiful example. During the last two to three million years, rhinos have changed little anatomically. They are the poster children of success on a continent where ecological conditions have been relatively stable. Even today, lions and hyenas rarely bother adult rhinos. Both sexes maneuver almost as well as horses, and each is armed with saber-like horns. Although vision is poor, their sense of smell enables the creation of a spatial memory that operates across a rhino landscape just as global positioning systems orient us today.

As little as a century ago, black rhinos numbered sixty thousand. Today, only five percent of that number remain. Only after horns became big business did it become clear that rhino defenses were ill-adapted. Descriptions of their behavior 150 years ago, even before the poaching epidemic of the twentieth century, differ little from today's behavior, a testimony to their lack of capacity. In his 1856 book *Lake Ngami, or Explorations and Discoveries in the Wilds of Southern Africa*, Charles Anderson described their aggressiveness and frightfully poor vision: "of unprovoked fury, rushing and charging with inconceivable fierceness, animals, stones, bushes—in short, any object that comes their way." What may have been adaptive 150 years ago is no longer. Rhinos failed to accommodate for the arrival of a new predator—one with two legs and guns—and they were rapidly massacred from Somalia and Sudan to Angola and South Africa.

To combat the slaughter, the governments of Namibia and Zimbabwe attempted a radical conservation tactic: dehorning their rhinos. The logic was straightforward—with no horn, the incentive to kill should disappear, for there would be nothing to harvest. During the early 1990s, I developed a project in Namibia to evaluate whether this plan would work.

Were horns weapons against predators, adornment for mating status, or just structures from the past that were now useless? After more than three years of data collection, I achieved results suggested that females needed horns to protect their calves from predation by spotted hyenas. Mothers without horns did not successfully rear calves. In Zimbabwe the results were equally dire. Poachers killed more than ninety percent of the hornless white rhinos, male or female, mother or not. Although the dehorning strategy had been worth a try, it was not successful.

In the process of watching black rhinos on Namibia's deserts and savannas, it became clear that their behavioral responses to threats differed between natural predators and potential poachers. Females were more sensitive to lions and spotted hyenas, regularly charging both. Each gender, however, ran from humans, females almost fifty percent further than males—about four miles on average. Although black rhinos are not equipped to deal with modern humans, they clearly know how to distinguish between types of possible predators, four legged and two legged.

Other sorts of human handiwork have unraveled what were once stable relationships between prey and predator in Africa. With thick regal horns and dramatically striped hides, bongos are the largest of the forest antelopes. Living in thickly vegetated equatorial mountains, males reach nine hundred pounds, females five hundred. The species is in serious trouble, both because of habitat loss and poaching. Kenya's Aberdare Mountains harbor perhaps one hundred animals. Fifty years ago, bongos were doing fine. Today, lions are considered their major nemesis.

Lions did not historically occupy these thickly forested mountains, perhaps because the past prey base was meager. They began arriving once they got in trouble with cattle elsewhere. As a result, the Kenya Wildlife Service transplanted them to the Aberdares, tropical highlands that host one of Africa's highest densities of spotted hyenas.

Just how the ecological dynamics unfolded among hyenas, lions, and bongos has never been clear, nor is it known whether bongos learned to avoid lions or if lions have been the primary drivers of the bongo's rapid demise. Perhaps lions displaced hyenas from kills, and hungry hyenas were therefore more directly responsible. Whatever the truth, lions have paid a large price. More than two hundred have been removed in hope of offering bongos a better future. The bottom line, of course, is that neither the bongo nor the apparently introduced lion has fared well in the Aberdares.

Native carnivores aside, domestic dogs can also be formidable predators. They eat prey as different as marine iguanas, waterfowl, and wildebeest. When prey are ill-prepared for novelty, effects can be striking as was clear by the actions of a single rogue dog reaped on New Zealand's most celebrated bird.

Kiwis are secretive, nocturnal, and have bristly feathers. Lacking the ability to fly, and living in forests and scrub, they are survivors of the era of moa. They are also rare and in trouble, because they cannot deal effectively with alien species. The efforts of that one kiwi-loving dog underscored the birds extreme vulnerability; he killed about five hundred of the fowl-sized relicts during a forty-day assault.

These relationships speak to the question of whether today's native faunas can survive the recent arrival of alien four-legged predators. Can prey change their behavior fast enough to assure their coexistence?

AUSTRALIA OFFERS A different but no less distressing test, the results of which offer a cautionary tale. Dingoes are beautiful, ginger-colored, feral dogs that have now become naturalized. They arrived courtesy of Asian seafarers 3,500 to 4,000 years ago, 45,000 years after Australia's original human colonists set foot ashore. By the time Europeans arrived, dingoes had spread across the continent.

In the 1930s, three species mentioned in earlier chapters—Tasmanian devils, Tasmanian tigers (also called wolves or thlyacines), and Tammar wallabies—survived only on offshore islands. Because of its nasty sheep-eating habit, the magnificently striped thylacine was hunted to extinction on Tasmania. Devils and Tammar wallabies still persist on offshore refuges.

The involvement of dingoes in the contemporary extinctions on the mainland remains inferential, as does the role of behavioral naiveté. Some uncertainty exists, since dingoes are social living and, as pack hunters, they may have obtained prey by reliance on tactics that differed from native marsupial carnivores. Kangaroos, like ungulates, also evolved under pressure from cursorial predators. None of these fleet marsupial predators now remain, so it is impossible to know whether the antipredator behavior of large kangaroos changed after the arrival of dingoes.

Although kangaroos are typically quick on their feet, biologist Susan Wright watched a solitary male eastern grey "roo" and a single dingo interact: "The kangaroo high-stood, watched the dingo, thumped his tale on the ground and twice kicked at and hopped toward the dingo . . . [which retreated] slightly from the kangaroo. . . . Of the total time [63 minutes] the dingo was in sight, 51% was spent within 5 m. . . . When the dingo left, the kangaroo . . . lay down."

Red foxes, catlike canids with retractable claws, were introduced between 1845 and 1870, and they offer a different insight. Their colonizing waves expanded outward from Melbourne at 75 miles per year. By 1930, they roamed half the continent. Red foxes, unlike dingoes, stalk prey rather than chase it down, and the foxes proved highly troublesome.

At least twenty medium-sized species of marsupial have gone extinct on the mainland during the past fifty years. Alien predators, which also include millions of feral cats, have been implicated, in part because of the inadequate abilities of prey to adjust. Ecological explanations are rarely so simple, and disease, grazing, competition, and changes in fire regimes are all likely contributors as well. Nevertheless, numerous marsupials have been noted for their abject lack of astuteness to new predators.

The Tammar wallabies of Garden Island are especially intriguing. Due to rising sea levels, they have been estranged from terrestrial predators for seven thousand years. In 1973, a bridge was built to the island, and in June 1996, a red fox followed the 2½ mile causeway to the predator-free wallabies. Eleven were killed in the first 48 hours, and a minimum of 25 died in 11 days. Although this had to be a case of surplus killing, since some bodies remained untouched, the issue at hand is whether naiveté altered whatever rules might have been inherent in the evolutionary game between prey and predator.

The eastern barred bandicoot, an elegantly snouted nighttime forager of insects, berries, and worms, is another species whose continental range has collapsed. Although critically endangered on the mainland, they remain abundant in Tasmania. Like this bandicoot, black flanked rock-wallabies have a similar distribution—occurring on at least seven southern and western offshore islands but rare on the mainland. Finally, there are burrowing bettongs. These squat, rabbit-sized, yellow-gray animals subsist on bulbs, seeds, and insects. They too have become extinct on the mainland but remain on islands off Australia's west coast. Each of these four species has had similar experiences with foxes.

Bettongs were introduced from Dorre Island onto the adjacent mainland. Upon first contact with foxes, 77 percent died, mostly in a spree that mirrored surplus killing. Three years passed until the next encounter, when 36 percent of the bettong population was removed, again as a surplus kill by foxes. For wallabies, an analogous pattern emerged, with 11 animals killed over just a few days. For bandicoots, it was essentially the same.

Also embroiled in Australia's nightmare are Eastern quolls, which offer one of the clearest cases of possible behavioral failures. This cat-sized nocturnal carnivore sports a spotted tawny coat and a long furry tail. It dines on insects, birds, and carrion. On the mainland, this quoll followed the path to oblivion, as did its fellow meat-eaters, disappearing by the 1960s. The deliberate introduction of about a dozen red foxes onto Tasmania prompted University of Tasmania biologist Menna Jones to ask whether the quolls were sufficiently adept to thwart foxes. She tested her ideas by playing the sounds of dangerous and neutral species. By all indications, Eastern quolls were not ready for foxes. Juveniles did not know how to respond, and adults were only slightly more alert. However, when exposed to species they evolved with, such as Tasmanian devils, quolls became more vigilant and reduced movements, a behavior that rendered them less likely to be eaten.

If naiveté is a fundamental problem, then behavioral modification might be part of a solution. That is the hope of a research team led by Ian Mclean, whose goal is to teach prey to fear introduced predators. Like bettongs, barred bandicoots, the black-flanked rock-wallabies, and others, the rufous hare-wallaby has become extinct on the mainland. The last individuals perished in the Tanami Desert during 1991 after a spate of simple bad luck—a single fox's handiwork coupled with a fire. Ian, who once worked with Canadian ground squirrels and grizzly bears, argued that if the rufous hare-wallaby could learn to fear alien predators, its chances for survival would improve. He conditioned captive wallabies to fear cats and foxes. However, because the models he used were larger in size than the five pound wallabies, it was uncertain whether the fear resulted from predators or was simply an aversion to larger objects. Promising, however, was the finding that levels of wariness increased after eight months, an indication that the capacity for learning exists.

As attractive as the behavioral naiveté hypothesis is to explain the collapse of medium-sized marsupials across Australia, other explanations are also possible. For instance, surplus killing is not restricted exclusively to interactions involving introduced predators and prey, as wolves and bears are known to kill multiple deer and caribou. Still, slaughters of naive marsupials, if sufficiently frequent, will easily lead to a population's decline.

Should there be any doubt about the role of ineffective behavior, Matthew Flinders provided exceptional observations when he reached Kangaroo Island in 1802. "On the first day the landing party killed 31 [western gray kangaroos] with such ease that some were simply knocked on the head." These animals had been separated from the mainland for at least 9,500 years and had lived with aboriginal people—though not for at least the past 2,250 years. Hence, they were not likely to remember the danger of humans.

By contrast, Captain James Cook found just the opposite. In 1772, he navigated his British vessel *Endeavor* to the Yorke Peninsula. His greyhounds and sharpshooters were unable to kill kangaroos, probably eastern grey or common wallaroos, for three weeks. Unlike the island, where Flinders' group beat a hasty end to the naive roos, those on the mainland presumably had ample experience with dingoes and aboriginal hunters.

If this were true, then humans and dingoes were responsible for the differences in how kangaroos responded to predators. However, such a generalization may be dangerous, because other aspects of environments inhabited by the kangaroos might have varied as well. In essence, we just do not know enough about the past history of predation on the mainland to make sense of the behavior.

It's easy to understand why skeptics dismiss evidence without supportive experiments. Yet, experiments over large land masses or spanning many years are rarely possible. From a conservation perspective, the issue of interest is not what happens upon first contact, or second or third. It centers on whether individuals learn and whether, as a consequence, chances for a population's survival increase over time.

AFTER THEIR SIXTY-YEAR absence, wolves returned to the Tetons late in 1997. Following on their tracks was local fervor over the expected crash of elk and moose. These worries were grounded in fact and reality. Predator-free populations typically have lower rates of mortality than those with carnivores, and they usually occur at higher densities. If high densities were a goal, then carnivores were anathema.

Through 1998, my data on radio-collared moose and their calves indicated high survival; in the year after first contact between wolves and calves, more than half died. Just as in Australia, the bounds of interaction between prey and predator remained tenuous. Speculation was rampant that moose would follow in the steps of their Pleistocene ancestors; a blitzkrieg lurked silently in the shadows.

Since moose had lived wolf-free for more than a half-century, no one knew if the return of wolves would be a death knell. Underpinning the idea that populations would plummet were two important assumptions: one behavioral, the other ecological.

The notion that moose were no longer adapted to wolves had intuitive appeal and some empirical support. Wyoming moose were very naive. My sound playbacks had demonstrated a lack of responsiveness to not only ravens and coyotes but also the howls and urine of wolves. If moose had forgotten their prior enemies and did not have the capacity to recall them or learn about them, their fates were predestined.

There was, however, one overwhelming difference between the collapse of Australia's fauna and the return of wolves to Wyoming. Moose and wolves had a long history together. From the cold woods stretching across northern Europe and Asia to the boreal tiers of North America, there were thousands of generations of coexistence. Had wolves been so abjectly destructive, moose would not have persisted. Their evolutionary pathway would have terminated at the end of a cul-de-sac where they would have joined mammoths and mastodons. Places must have existed where moose were initially unfamiliar with wolves but survived beyond the first contact.

Indeed, at least two such regions exist. The first, Isle Royale, is a small island in the cold waters of Lake Superior. Its moose population is the most studied in the world, as Rolf Peterson, his colleagues at Michigan Technological University, and their predecessors have concentrated efforts there for half a century. Moose colonized the 225-square-mile island just after 1900. They remained predator-free until wolves arrived in 1949. So, like my study animals on Alaska's Kalgin Island who remained predator-free for fifty years, those on Isle Royale also lived without dangerous carnivores for a similar period.

If ten to fifteen generations of abject naiveté destined moose to be highly vulnerable meat-on-the-hoof, we would expect to see a severe population crash coincide with the arrival of wolves. That did not happen. The big browsers of Isle Royale persist today, their fluctuations explained by a combination of predation, weather events, and food supplies. Despite the fifty-year blip in experience with wolves, these moose either redeveloped ways to deal with them or never forgot. Unlike arctic ground squirrels, pronghorn, and other mammals exposed to relaxed predation for thousands of generations, moose have apparently maintained the ability to respond within just a few.

Scandinavia is a second area where modern moose once again cross paths with another predator—brown bears. Along the Norwegian-Swedish border, bears were driven to near extinction more than a hundred years ago. The range of bears is now increasing there just as it is in Yellowstone. Along the colonizing front, bears encounter moose that have previously been naive. If naiveté truly renders moose vulnerable, then those on the front lines should be more likely to succumb than moose that have lived alongside bears at the center of the range. To appreciate this experiment requires a step backward in time.

One February, I found myself at a conference in Hell, the name of a town on the coast in Norway. At the same scientific gathering was a tall, soft-spoken American who is now a professor at the Norwegian University of Life Sciences. Jon Swenson was unsure if my speculations about bears and moose learning abilities would prove true. His long-term study of moose in Scandinavia held possible answers.

Six months after our meeting, Jon sent a message from Norway. Scandinavian moose were less successful at avoiding predation by brown bears along the colonizing front than were their kin who had been living at the core of the bear's range. This information was derived by Jon and his co-worker, Inga Lill-Persson from analyses of 23 encounters between moose and radio-collared brown bears. Despite the limited sample, naive moose—those where bears were just arriving—were about six times more likely to be eaten than moose who had coexisted with bears.

As in the American West, rural Scandinavians are concerned about predator expansion and the push for protection of bears and wolves. The issue is as controversial there as it is in the US, even though ecological conditions vary greatly between the two regions. The expectation of a collapse in moose numbers in northwest Wyoming is predicated on the belief that moose will fuel the diets of wolves. The assumption rests on shaky ground.

In the Yellowstone ecosystem, nearly a hundred thousand elk coexist with only several thousand moose. If I were a hungry wolf faced with a choice between subsisting on something big and dangerous or medium-sized and less lethal, I'd pick the smaller quarry. While wolves and other animals may not make decisions about foraging with much forethought and choices like these are overly simplistic, carnivore decisions are driven by factors that affect hunting success—snow depth, habitats, hunger, and the caloric value of prey. Economics govern where and when, what and how to hunt. Prey abundance and size, amount of meat, and risk of injury all factor into the equation. With thousands of elk in the Teton region, it does not make overwhelming sense to expect moose to be preferred over elk. Then again, if moose remain unwary and are easily dispatched by wolves, their future could be bleak.

IT WAS EARLY May on the eastern edge of Grand Teton National Park. Gentle rays of a midday sun warmed the air. Sage dappled the slopes in greens and grays. Balsamroot painted the landscape yellow. I watched elk in the meadows below. During the prior two years, the Teton wolf pack had denned in the same location, not more than half a mile from where the elk now grazed.

"Stand your ground" is not a mantra known by elk or moose. Neither species, however, are as vulnerable to predation by wolves if they remain as they are when they flee. In the half-century of research on Isle Royale, moose that have stood their ground have avoided being killed. Those that have run have been less fortunate, a behavior that also characterizes the elk of Grand Teton and Yellowstone.

A solitary black wolf appeared but stopped abruptly. It was the pack's dominant male. The elk sensed something. They became rigid and gazed.

The few on the outskirts of the group moved inward. The wolf remained, patiently lowering its body to ground level. Its ears were flattened back.

The elk were nervous, their heads held high. Ears whirled like blades on a helicopter. Several animals began to shuffle. Others followed. Only a couple of central individuals still fed. The group numbered 23.

The wolf darted out, its pace accelerating. The elk fled, but two lingered ever so slightly. Soon, the wolf was on the heels of a young bull. He turned sharply, powering ahead, easily outdistancing the wolf. The chase continued. A female faltered. The wolf was at her hind legs. She tripped and rolled, but sprang up. The wolf lunged, this time catching her on the flank. Again she stumbled. This time, she did not rise.

As the dust cleared, the wolf's powerful jaws clamped like a vice grip on her throat. The other elk kept running, but suddenly stopped two hundred yards away. Within moments the wolf had subdued the unlucky cow. Perhaps the group knew the chase had ended. He fed on her for six minutes, before moving off 75 yards. Then, he curled in the warmth of the sun, and slept. Seven minutes later, a howl from across the valley punctured the silence. The wolf heard it and stood.

The elk, now numbering 22 animals, had gone back to nibbling on lush grass. When they heard the distant howl, the majority broke off their feeding. Evidently, they were using their sensory prowess to ingest the sounds, smells, and sights of their enemy. Future survival would depend not only on their immediate ability to detect danger, but also on how effectively they avoided it.

I expected the wolf to return to the carcass. He did not. Instead, a new stalk was underway. The elk again became wary. The air was calm, the sun still shining brightly. Time appeared to stop. I sat cross-legged and readjusted my scope with one hand and padded along the ground to feel for my binoculars.

The wolf began to make his move. He shifted in and out of view, hidden by thickets of sagebrush. He moved with stealth, with purpose. A grassy swath some fifty yards across now separated him from anxious elk. They stood, bunched and staring. None fed.

I switched to binoculars. The standoff lasted about twenty seconds. The male bolted forward. Within two minutes, another elk, again an adult female, stumbled. He killed her, also by the throat. This time he fed for only four minutes.

I marveled at what I had just seen. A single Teton wolf killed two adult elk 18 minutes apart. Wolves were supposed to eat what they killed. That had not occurred. Were these adults really just as vulnerable as nascent calves? Perhaps so. Perhaps the wolf was just practicing skills it might later need. To the north in Yellowstone Park, at least nine wolves have been killed while hunting elk; one by moose, the other by bison. Predation is a dangerous business. Real world biology is complex, adaptive explanations easily dismissed. Why had the wolf killed two elk?

Elk might benefit from these sorts of experiences. Those that died in the hunt obviously did not, nor did they have a chance to apply new knowledge to future situations. But, for others, the playing field may change. Perhaps, elk that had watched, been chased, or felt wolf teeth on their heels learned. If so, others might benefit in the future. Social learning was one mechanism by which prey might acquire skills to avoid predators.

While I did not know elk as individuals as I did my radio-collared moose, it was becoming clear that Teton elk were learning about wolves. Based on my playback experiments, I discovered that elk in areas of wolves became very wary, grouped, and ran far more quickly and frequently than elk from areas where wolves did not occur or had only recently arrived. Unlike moose that lived most of their lives solitary, elk had access to advantages that accompanied life in herds.

IN YELLOWSTONE AND Teton parks, winter arrived on the heels of a summer with less rain and warmer temperatures than normal. If ungulates could communicate with humans, they would have revealed an impending expectation for the worst. The crop of young elk and young moose born eight months earlier in May and June were ordained for trouble. Through no fault of their own, they entered a world in a year when summer was dry.

Their access to plants that otherwise should have been rich in nitrogen was limited. Their mother's milk should have been thick with fat and protein, but it too would be less nutritious. By winter, all animals—males, females, juveniles—would be in poor body condition. If winter was anything other than mild, high mortality was likely. This, coupled with the suspicion that wolves were lurking around every corner, would send an ominous message to those rooting for a banner year for survival. The relative number of recruits to make it past winter might be down. Wolves would garner the full brunt of the blame.

Over the next few years, moose did die. Some perished from circumstances that were far from natural. Several became entwined in fences and struggled for days before succumbing. Cars took others. Hunters shot a few.

Natural deaths, however, were the rule, and some were baffling. In

May, several pregnant near-term females died. Of four I found, all were recumbent. Three formed the center of a wide circle where the vegetation had mysteriously been beaten down. It was as if the moose were the hub of a bicycle wheel, with the rubber instead of the vegetation removed from the tire. In all four cases, the anus had been opened and tissues removed. Puncture wounds to legs, flanks, and throats were lacking. Pleural and thoracic cavities were untouched. Was this the work of a predator or something else?

While flying her Cessna 182, Lighthawk Lisa Robertson reported wolves at a carcass up in the Mount Leidy Highlands. That was all that Doug Brimeyer, a biologist with the Wyoming Game and Fish Department, needed to hear. Wolves were clearly the problem, especially since moose rarely died in late May. My assistant, Noah Weber, and I were less sure, even though there no denying the wolf tracks.

We carefully sliced the skin on the moose's face and throat. We worked our way down the center and across to its rear legs. There was no evidence of hemorrhaging. We found several bite marks on a rear leg, but these had occurred after the animal died. Some steak had been removed from the animal's back. When we looked beyond the corpse, we found drag marks and blood. A small object had been removed. A calf.

At three of the four sites, we found either the hooves of a fully developed fetus or tiny deciduous teeth. The body parts coupled with the puzzling circles in vegetation helped solve the mystery. The mothers had kicked furiously while spinning on their side as they died from birth-related complications.

In the absence of our investigation, it would have been easy to conclude predators were responsible. Two items now suggested otherwise. The first concerned the lack of evidence of predation. The second was the locations of sites. Mothers give birth in closed, thickly vegetated areas, not in the open as, such as the sites where the dead mothers were found. I speculated that mothers who were experiencing birth complications may have run, and just happened to drop down out in the open. I didn't really know, nor did the veterinarians that I consulted. Fieldwork is like that; some secrets remain with the dead.

Dear Mom and Pop, Feb. 15

You asked the other night what it is that I do all day, and how it is that I can like winter when you are enjoying California. Here's a glimpse of last week.

Tom Roffe, who you may recall is a veterinarian, and I went to put collars on moose. It was minus 27 degrees this morning. We trudged through 3 feet of snow. We hit a few snow bridges over what we hoped was a frozen creek. When we tested it, we discovered the bridge weak and that water was running below. We hoped the bridge would hold us (it did). Our goal was to find a specific female that Kim had named Pacifica because she lives along Pacific Creek. We wanted to check on her body condition, to see if she is pregnant, and, most importantly, her old collar had old batteries which needed to be replaced.

Tom is a great guy, and we always chuckle and make fun of each other. This time, as we were going up and down slopes on snowshoes, we kept asking ourselves why we do this. Tom just turned 50 also, and we are just a few days apart, so we give each other severe grief. His hair is not graying like mine, but I point out that he cannot grow his long. Anyway, you get the idea about our banter and relationship. More seriously though, we do ask why we are chasing moose (I mean really—what a life?) when others at our stage think of retirement, or the Caribbean, or anything else other than exertion, sweat, labor, the salted sting of sweat as it rolls into their eyes.

A few days later, I left home from Teton Valley at 5:40 in the morning. It was still, dark, and clear. At the base of Fox Creek it was relatively warm, 4 degrees. As I passed through Jackson, the thermometer read minus 14. I continued north, passing the refuge where it was now getting light. There were some 7,000 elk. I was heading north about 50 more miles to check the status of and photograph the twin calves of a female named Nathalie. Yes mother, she was named for you. Nathalie is a beautiful cow, just as you are still a beautiful mom. I hope it doesn't offend you to have a moose named in your honor. She really is a sweetie.

Twenty miles up the road, I stopped to check the signals of a few other animals. Then, I realized my jacket and gloves were 70 miles away (at home). That's your brilliant son. The sliding door on the van wouldn't seal shut, so I gave up after my 12th try. I used bungee cords to keep it partially closed.

As I crossed the Snake River, ice chunks floated by. I parked next to a snow machine and pulled out all my remaining clothes, and began a short 200-yard walk to a higher spot so that I could scan for Nathalie. Half-way there, my nose was killing me, and the hair inside was not happy either. At this point I was thinking I really need my thickest down mittens and my face needs real protection. Then, your middle son had an epiphany; his mind began to work. I remembered the spare trunk in the van and the seriously cold weather gear inside. I began to run back, but my face became so cold all I could do is walk. At the trunk, I found a face mask, as well as my rabbit fur-lined hat, and an additional balaclava. Then, another face mask appeared. This one was a good friend, the one I used in Svalbard.

I then crossed a frozen lake on snowshoes. Finally, my body was heating up even through the actual temperature was minus 34 F. After about an hour, I found Nathalie and her two calves.

On my way back, I checked on another animal, Willow was her name. She just died, and I needed to get her jaw. Only 3 days earlier she was alive. By the way, most of the animals are living; when they die it is more exciting (plus you asked for my week.) The problem was that I was supposed to pick up Sonja after school, but I was still too far for my cell phone to reach Carol. So, I drove another 15 miles to make the call, and then came back for Willow. Coyotes had just found her frozen body, and two ravens were waiting.

I got home after dark, and will see Soni this weekend. Sorry if this was a bit of a ramble. Not all days are field days, and sometimes I just work on data. On others there are meetings with government or with ranchers or environmental groups.

I hope this helps, but if not come back out for a visit. I know your choice would not be winter, which is fine with us. Otherwise, we'll all see you this summer. Sonja is anxious to try surfing. Am hoping the two of you are well. Kim sends her love.

I'll call sometime soon.
Love you.

MOOSE LIVE FAR more solitary lives than elk, and opportunities for social learning are inevitably limited. Mothers who die obviously cannot pass their knowledge on to newborns. Offspring, however, have opportunities to learn when their mothers survive. Young moose develop similar migration patterns as their mothers, adopt home ranges in the same general area, select comparable habitats, and often mimic foods their mothers eat.

Whether such behavior develops by trial and error or by copying is not entirely clear. In part, it must stem from association and common travel since offspring who survive beyond the immediate death of their mothers are less likely to share in the same habits as those whose mothers live.

The death of a mother or her offspring creates potent opportunities to examine how individuals modify their behavior. Since such events are infrequent, each alone represents nothing more than a mere anecdotal account. When anecdotes are weaved together however, a picture emerges about how behavior changes after an experience with danger or death.

The first hint of something unusual surfaced on a dark wintry evening. Author and adventurer Ted Kerasote drove his little Subaru down the Gros Ventre River road. The hour was late, the wind blowing snow, ice thick, and visibility near zero. Two moose stepped out. One wore a radio collar—Daisy.

Ted hit the brakes but not in time. He slid into Daisy. She ran off, followed by her seven-month-old calf. The next day I expected to find a wounded or dead Daisy. During the four years I had known and studied her, she had never left the area. I listened for her, and searched her usual haunts—willow thickets and cottonwood forests. Her total home range size was only about four square miles. There was no signal.

I expanded my search. I went to more distant areas and traipsed through them again and again. I inched up hilly knobs on snowshoes hoping to detect a signal. Still nothing. I called Mountain Air Research to arrange a telemetry flight. The weather was to be snowy for the next few days, so flying was out.

I drove east, past thickets of aspen and beyond sandstone cliffs. In the Gros Ventre River drainage, well beyond the park, I still found nothing. After more miles a faint signal pulsed. Daisy was alive. I stopped and hiked, finally finding her, although with quite the limp. Alongside was her calf.

Daisy had moved ten miles through deep snow and unfamiliar terrain. Her movements followed what had to have been a very bad experience. Daisy hung on for a few months but died later that winter, a victim of malnourishment. Two weeks later, her calf died.

Had Daisy been the sole animal to undertake such an unusual movement, I'd likely not have given much thought to the behavior of animals after a stressful interaction. Over time, however, a different sort of snapshot was developing. The behavior of female moose would surprise me.

One female, Lola, moved five miles immediately after her calf, Lou, was killed by a car. Then, there was Tae. She moved eight miles after her calf disappeared. Although I didn't know why hers had vanished nor why Tae had moved, mothers with surviving calves rarely moved long distances.

They did not go up and down mountains or over terrain unfamiliar to them. Mothers who had not lost calves showed great fidelity to their birth areas year after year. Those who lost calves were about six times more likely to shift to new sites in the following year.

My Alaskan colleague, Ward Testa, published a paper that shed new light on moose movements. Like me, Ward found that successful mothers returned to the same region. With his larger sample, Ward noted that mothers who had lost calves gave birth about six miles away the following year. Mothers were modifying their behavior, perhaps in a fashion that reflected they knew something about predators. These mothers had to be cognizant of their local environs.

IN 1999, TWO EVENTS would change my mind about Wyoming moose and their ability to learn. The first stemmed from the hundreds of playbacks I had conducted. Mothers that had not lost calves were unremitting in their unwariness. They still ignored virtually all sounds. These were not just the moose habituated to humans, for those occurring far from roads also showed disinterest in the auditory cues, whether raven or tiger, howler monkey or wolf.

Mothers that had recently lost offspring differed in two important respects. If calves were lost to wolves, mothers responded to wolf howls with hypervigilance. Not only was foraging terminated, the mothers also became nervous. They walked away, departing the area instantly. Conversely, mothers whose offspring died from starvation or vehicle crashes still ignored the sounds of wolves, just as did mothers whose offspring survived.

This behavior suggested that it was not the act of losing a calf, per se, but rather the involvement of wolves themselves that had prompted the maternal change. I did not know how mothers made the association. It seemed unlikely that they developed the behavior the way that elk might—through observational learning. There must have been a complex process that enabled individuals to associate howls with wolves and deaths of their calves. This seems remarkable, because wolves do not howl during active hunts, although they might do so as the pack coalesces prior to one.

The second event concerned the other major predator in my study area, grizzly bears. About a dozen adult moose had already been preyed upon, so these females were no longer available to transfer knowledge, if in fact that was how other moose learned about bears. Two items made me think otherwise.

The first concerned the moose from Alaska's Denali and Talkeetna regions. There, animals were highly responsive to bear scat. Additionally, Teton moose from areas with high grizzly bear densities reacted more strongly to bear odors when they had lost calves. I did not know if bears had preyed on the missing calves, but the behavioral evidence certainly suggested that calf-less mothers had a stronger aversion to bear odor than mothers whose calves survived. Just as in Scandinavia, moose living with bruins were expected to be savvier if they had coexisted with them for some time.

The second insight came about as a consequence of the distribution of moose. Mothers, I predicted, would choose habitats differently than non-pregnant females or those who had aborted their fetuses. The rationale was that mothers should minimize chances of predation by grizzly bears while seeking areas that offered nutritious food.

The variation in habitat choices could take different forms, especially since lactating mothers have needs that differ from those not producing milk rich in protein. Some sites might provide better access to water or different cover. Some might be islands or spruce forests, or deadfall. But the defining factor had to be the probability of encountering a grizzly if bear predation was guiding a mother's options.

To test these ideas, I examined the locations and habitats used by all radio-collared moose in late May and early June, the period when ninety percent of the births occurred within nine days. More than seventy percent of the pregnant cows gave birth within five hundred yards of paved roads. One was under a bridge, another next to a culvert. Several of the others had migrated to remote locations near Yellowstone Park's southern boundary. Gypsy, named for her habit of moving long distances, was 15 miles from the nearest road. She was in an area where grizzly bear scat was at densities almost as thick as flies. She was near sites where two people had been mauled. However, Gyspy, too, seemed wise, and during three consecutive years gave birth on ridgelines at 8,600 to 9,000 feet—where June's snow was still deep and bears unlikely to hunt reddish-brown skinny calves.

By contrast, non-pregnant females or those who had lost their offspring in utero selected habitats that averaged 2½ miles from pavement, and only one was regularly located near a highway. The distributions of female moose, when judged by reproductive status, were really different. Most interesting, perhaps, was that the geography of births had changed during the ten years of my project. Mothers moved progressively closer to paved roads on the day of birth, a shift that coincided with the expanding range of grizzly bears.

The endangered bears are generally averse to highways, usually spending most time more than five hundred yards from pavement. I didn't know whether my data reflected a change based purely on chance—that moose mothers had given birth in areas not frequented by grizzly bears—but I suspected otherwise. Mothers were not using roads to enhance access to minerals, because Wyoming roads were not salted, even in winter. Similarly, food was not a likely factor, since riparian areas were not distributed primarily near roads, and the riparian habitats had not changed over time. Most critically, mothers in the southern end of my study regions, where bear densities were lowest, did not show the propensity to change toward as had mothers from the north. It was as if the pregnant females from areas of greater bear densities were intent on reducing their calf's chances of becoming bear bait by their increasing affinity for roadside births.

One day in late May, I stopped to check the status of a new calf, only a few hours old and three hundred yards from Teton Park's busiest highway. As soon as I dropped off the pavement, the willows along the Buffalo Fork River grew thick and a brown muddy torrent, swollen from snowmelt, streamed past. The signal from the radio collar pinged loudly; the mother was close by. I looked for a tree just in case I was rushed. I held an antenna in one and pepper spray in the other.

Feeling vulnerable when I could not see, I continued moving forward, the pinging telling me the gap was narrowing. It was good to know that I was still alive, hypersensitive myself. With each step, I ducked, turned, and stood tall while trying to be silent. Finally, I neared a tall cottonwood. It had no branches for climbing but somehow I felt better.

The moose appeared. She was alive; actually too alive. Engorged with rage, her ears retracted, her nape fur fully erect. She ran back and forth, collapsing vegetation in her path. In a fleeting moment, I saw the calf.

My heart beat wildly. I could not understand why she wasn't focused on me. I glanced sideways. And then I knew.

I stood between her and another object, a large shape less than fifty yards off. Tall as my hips and dirty brown, it, too, was mobile. The mother cared nothing about me. I was a fly, a parasite, a mere annoyance.

The object had grace, power. Simultaneously, it captivated the moose and me. It was a grizzly. I had walked smack into the center of an encounter, and now stood between an angry moose and a hungry bear.

Immediately, the bear swiveled. It turned from the moose and did a reverse charge. I had no time to think, to react. In its path was water. The river exploded as the bear drenched itself, cutting a swath through the muddy current. Left behind were large six-inch-wide tracks. My eyes transfixed on the spot where the bear had disappeared. The vegetation

FIGURE 31. On Svalbard, an unhappy Kim warms up next to the heater that required more than two hours just to heat coffee.

FIGURE 32. The joy of morning coffee.

FIGURE 33. Windblown tracks of reindeer (small) and polar bear (large).

FIGURE 34. Snow machine, playback gear, and wild reindeer.

FIGURE 35. Mother and daughter and linear dimensions used to estimate head size (photo taken at a distance of 16 meters from the calf).

FIGURE 36. *Ger*, Gobi Desert, and Tsu-Tay Glacier in Altay Mountains in western Mongolia.

FIGURE 37. Nomadic Mongolian herders migrate near the Altay Mountains.

FIGURE 38. Saiga are found only in central Asia, though they lived in Alaska 30,000 years ago. This female is from Kalmykia, Russia. (Photo by R. Reading.)

FIGURE 39. The Pallas cat crouches in preference to running.

FIGURE 40. Death assemblage of khulons, gazelles, and camels near Madahk.

FIGURE 41. Takhi (Przewalski horses) in Hustai Reserve, Mongolia.

FIGURE 42. Girls peering at our passing convoy.

FIGURE 43. Vicuñas in the San Guillermo Biosphere Reserve in the Andes.

FIGURE 44. A cougar approaches in winter. (Photo by T. Ruth.)

FIGURE 45. Andean mountain cat. (Drawing by Sonja Berger.)

was thick. It was time to leave. I glanced at the mother, enjoying my departing look.

From the putative safety of the pavement, I wondered what moose might now adopt in their bear-avoidance repertoire. Bears were learning where to find baby moose, even near the busy highway. I knew that moose mothers also learned, and that their behavior was flexible. The system had changed greatly since I first arrived 13 years earlier. In 2007, at least eight grizzly bears were using areas in and around highways; a person was mauled less than six hundred yards from a tourist lodge, and somewhere around fifty grizzlies used some portion of Grand Teton National Park.

In this rapidly changing dynamic, I was puzzled about who learned faster, moose or bears, and whether learning about predators ultimately led to greater fitness. Some evidence pointed to an obvious answer. Moose mothers in the Tetons whose calves were killed by predators developed greater vigor in their vigilance when hearing wolves or smelling grizzly bears than those who had not. Also, as with mothers in the Tetons, those from the Talkeetnas and Denali who had not lost calves in a given year repeatedly used the same areas to bear offspring in the next year. Those who were unsuccessful due to predation shifted birth sites in the next year, on average about six miles. Hence, I knew mothers do things differently when experiencing both ends of the predator spectrum—success and failure.

The question of fitness, however, is tricky, because fitness is not merely reproduction in a given year. It is the relative contribution of genes as propagated by offspring over one's lifetime. While it is intuitively appealing to think that learning is related to fitness, few studies have been capable of addressing such issues with respect to mammals and their antipredator behavior. From a population perspective however, it is gratifying to know that behavior is malleable and changes can occur in adaptive ways in short time frames, at least in response to native predators. While bears and moose share a long history, the rules of fair play are regularly altered by us, in this case, via highways. Moose simply used our paved handiwork as a form of human shield against predation by bears.

HUMANKIND HAS BEEN fascinated with wildlife ever since we first scavenged and hunted. Over time, our tactics have changed. We fashioned sticks into spears, bones into clubs and awls, and fluted obsidian into projectiles. We dug holes and developed snares. We constructed visual barriers and drove animals over cliffs. The techniques were modified by trial and error, failure and success. They took time to perfect.

The rules, if any, differ today. No longer do large animals have sufficient time to adjust their biology. Beyond our "modern" techniques for hunting—GPS units, increasingly powerful rifles, and machines that take us further into the back country—even nonhunted animals are desperately out of synch with our world. In just a few generations, habitats are erased, rivers polluted, air poisoned, soil defiled. Precipitation and temperature regimes fluctuate wildly, and the planet grows warmer. Even if the behavior of a fortunate few manages to shift from innocent to savvy, there are issues other than alien predators. Habituation is one, a behavior that transcends into de facto trust.

Betrayal comes in ominous forms. Among the most disheartening may be modern deception by poison. Depending upon the substance, an animal's neurons fire uncontrollably. Breathing is paralyzed. Vomiting and diarrhea become explosive. Vision blurs. Death is not quick.

The victims of Jackson Hole's 2003 poisoning epidemic knew this well. The toxic substance Temek was carefully injected into hot dogs and then tossed from passing cars by some deranged motorist. Ostensibly destined for wolves, it killed more than a quarter of the park's radio-collared coyotes. Eagles and ravens died, as did at least two moose. Nearly a dozen unsuspecting pet dogs were also sickened.

Betrayals are especially harsh when an animal loses its fear of humans. No longer does it flinch, stare, or run as we move about. We meld as nonthreatening beings into their environment. We are accepted, and that is where the duplicity begins.

When our flower gardens and shrubbery disappear into ravenous guts, human tolerance for big browsers wanes. Beyond the suburbs, farmers and ranchers lose money when their alfalfa is massacred. Villagers in Africa and Asia lose entire livelihoods when elephants visit. Controlling the beast becomes a moral conundrum. What does it take to kill a docile animal feeding peacefully with no fear of an armed human?

The strokes of a pen, unscrupulous behavior, and war have all led to slaughters. In 1992, the eastern third of Canada's Algonquin Park was opened to the First Nations people of Golden Lake. More than one hundred habituated moose were slain. In Angola, the story differed, but only in detail. Its national emblem, the stately Giant sable antelope, had been deliberately habituated to vehicles by researchers. The practice might have worked well for data and for subsequent tourism had it not been for a 25-year civil war. Unsuspecting animals were mowed down by soldiers. Whether in reserves, near human dwellings, or places far from people, habituated animals do become easy targets. Shooting them in the absence of fair chase corrupts a bond we share, one no different than shooting a dog. While controls on population levels may be understandable, a more pertinent issue is how best to do so without betraying a silent trust.

THE FATES OF countless animals remain in our hands. Beyond wild species are domestics, those that we have modified for one reason—our needs. They produce meat, milk, fiber. They carry our burdens. They nourish our souls.

Nowhere are our actions as cogent as they are with species we immortalize or despise. Horses are loved, except where feral. Zebras are generally the opposite—loved by wildlife enthusiasts but ostracized because they compete with livestock. Dogs and cats engage the same genre. As pets they are either loathed or loved. Wolves strike fear, lions respect. Whether domestic or wild, friend or family, what do we do when the ones we admire grow old? Do we change the rule to which they have become accustomed? Is there really a difference between killing a wild animal and a domestic one? How do we say farewell to a friend?

SAGE HAD A small head and erect ears. Otherwise known as a Queensland blue heeler—a mix of dingo, collie, and Dalmatian—she had come from a ranch in central Nevada. Over time, she would help find moose. She would lick Sonja's face, starting with the day she arrived home from the hospital. Sage slept at my feet when I analyzed data, kept me company in the field, and crawled into my sleeping bag with me on wintry nights. Our bond was special.

I remembered when she used to balance bones on her nose and when she tried to round up bison. I recalled the moose that nearly killed me just because Sage had discovered her newborn, and the grizzly she enticed for a visit. Like with all of us, age crept up ever so silently. At 13, Sage stopped jogging with me. At 15, I had to carry her up the stairs.

On September 7, 2001, Kim came home and again asked whether Sage was hungry. Having not eaten for nine days, she had shed thirty percent of her weight. We knew her time was coming. Sonja offered a final good bye. Sage managed a tail wag. Two days later and wiping away tears, Kim said her good-byes.

Just the two of us, Sage and I drove up a dirt road. The air was clear, the sun cresting the eastern ridges. Sage looked at me, knowing we were going somewhere good. I patted her head, "good girl." Then, I made a raven-like "caw," and her ears pricked up. It was still cold when I picked the spot. Sage climbed out of the van. I checked the .357 Magnum. A few flowers still bloomed in the meadow.

I waited for Sage to rest, not knowing if I had the courage. Sage walked to a bush and did the typical doggy thing, she smelled it. I called to her. She stumbled back. Liquid oozed from her butt. She lay down, curling in the warmth of the sun. I removed the pistol.

Her eyes opened. I ran my fingers up her nose to between her eyes. I inserted two bullets and placed the gun behind her ear. She opened her eyes, and closed them. I put cotton in my ears. The sun grew higher. Sage breathed deeply. I put my finger on the trigger, and gently squeezed.

Blood exploded. Her legs kicked violently. Her tail swayed. I looked away.

There was no burial, just rocks with blood. I picked up three stones—for Kim, for Sonja, and for me. I built a small rock cairn, dousing it with Sage's blood at the top. At the van, I looked at a small photo of a puppy and a 36 year-old man playing together. One had a wagging tail, the other a smile.

Two days later my own troubles paled in comparison. It was September 11, 2001. Fear struck the heart of American innocence. The rules of engagement had changed.

chapter 13

Nomads of the Gobi

[T]hough food is scarce, they [immense herds] have no fear of encountering their worst enemy, man; and far removed from his blood thirsty pursuit, they live in peace and liberty.

NIKOLAI PREJEVALSKY (1876)

LESS THAN TEN years after Russian-born Colonel Prejevalsky trekked across central Asia to reach the Far East's Sikhote-Alin Mountains, Swedish explorer Sven Hedin began his own epochal odyssey. It would span three decades, cross the roof of a continent, and descend into some of the planet's least friendly deserts. Both men appreciated solace and the splendor of distant lands where animals had no fear. In his 1905 account, *Central Asia; North and East Tibet,* Hedin remarked, "Of human beings we saw not the slightest trace, while the wild animals . . . were so little shy and so numerous that, I concluded, we were still a long way from the first of the nomads."

Although Hedin referred to wild yaks and kulans—the latter an equid known for its half-ass, half-horselike-qualities—central Asian animals have long contended with nomadic humans, other predators, and the vicissitudes of the land itself. Some had apparently perfected survival, for, as Hedin noted, "the wild camel . . . can scent a man at . . . 12 miles, and then he flees like the wind"; also, that camels live "as a nomad, for in the winter he generally goes down into the deserts of the lowlands, the Desert of Lop, and the Desert of the Gobi." The words linger, tributes to the past.

Today few camels persist, even in Mongolia where lands remain unfenced and open. As Asia's seventh-largest country and with only 2½ million humans, 99 percent of the surface is unoccupied by people or agriculture. Nearly half of the population still relies on traditional beasts of burden—horses and camels and a few yak. To survive this challenging part of the world, brightly dressed herders parallel the tactics of wildlife. They move, living as nomads. Relying mainly on *gers* ("yurts" in Russian)—tents constructed from canvas, wool, and wood—they shift their stock across the sweeping grasslands, settling where pasturage is good until it is time to move again. Doing similarly are Mongolia's gazelles, camels, and kulans, species that Hedin watched mixing together in large herds.

Ever so slowly, change permeates the steppes. A few motorcycle tracks have become many. Trucks replace horses. Habitat loss, desertification, and poaching become serious. Large international firms dangle hopes of sustainable economies steeped in mining or petroleum to unsuspecting dwellers of fragile deserts. In winter, the capital city of Ulaanbaatar sits under a veil of turbid clouds poisoned by petrochemical emissions.

The megafauna has suffered. Wild horses, known to Mongolians as *takhi* but to westerners as Przewalski (named for Colonel Prejevalsky) horses, were gone by the 1960s. The survival of the entire species once hinged on the fate of 13 individuals that were brought to zoos. Camels have fared slightly better. Bactrians, the two-humped wild species, are endangered but hang on in small portions of western China and Mongolia. Dromedaries, the single-humped version, survive only as domestics; the last of their wild relatives became extinct at least two thousand years ago. Mongolian tigers followed the same path. Like *takhi,* they disappeared years ago. Wolves remain the traditional bane and are persecuted almost everywhere, including protected parks.

I hoped to expand my studies into Mongolia. Some ten years earlier, George Schaller reported wolves were a primary predator of young camels, thus offering what I expected to be a reasonable basis to understand camel behavior in relation to wolves. However, given the earlier descriptions by Hedin, Prejevalsky, Morden, and others, it was already clear that wildlife harvests in central Asia were not sustainable. Three American biologists and their Mongolian colleagues had already confirmed high levels of poaching were decimating wildlife. Between 1995 and 2005, red deer populations dropped ninety percent; even marmots were being slaughtered for food and hide. Unlike my work from other areas where people were less of an issue, humans would have to figure prominently in any study of prey and predator here. Even where predators were locally extinct, prey might be responding to their replacement—humans.

Despite the loss of tigers, which were not abundant in Mongolia in the first place, other carnivores remain. The luxuriantly furred snow leopard persists in the Altai Range in the west, in the Gobi-Gurvansaikhan Park in the south, and in isolated desert mountains. The Eurasian lynx is in a few areas of the steppes and is well-known from traditional forested habitats. Wolves are widespread. Brown bears, though declining, receive conservation scrutiny, especially the thirty or forty that cling to a precarious existence in the Gobi.

Some protection remains in place. In the thirteenth century, Marco Polo reported, "Mongolians prohibited hunting during the birth and weaning period of hare, elk, roe deer, gazelle, and other animals. . . . Therefore there are plenty of animals and wonderful opportunity for the increase of animals." More recently, Mongolia committed 11½ percent of its land to protection. *Takhi* were reintroduced from western zoos into Hustai National Park, a hundred-thousand-acre preserve at the interface of grasslands and forested mountains. These striking dun and golden-colored horses now number more than 140 in Hustai and have been reintroduced at two other sites.

Among other interesting hoofed mammals are two species of gazelles, a wild sheep much like bighorns called argali, and a member of the goat family, the ibex. Five species of deer—elk, moose, roe, reindeer, and the diminutive tusked musk deer—occupy the Hentiyn Nuruu, a range of sacred, forested mountains north of Ulaanbaatar. Also occurring in Mongolia are wild boar and a thousand or so saiga, the world's northernmost antelope. About half the size of a pronghorn and weighing only fifty to sixty pounds, this bizarre antelope has a dangling proboscis and can attain speeds up to 45 miles per hour. Saiga inhabit dry steppes and deserts and are critically endangered.

My primary focus would be on a different species, one that I knew only from books, the kulan. Fleet but stout, tawny with powerful legs and beige rumps, kulans have short, erect manes, dark dorsal stripes, and thin fluted tails with black tips. These half-ass, half-horse animals are legendary for their endurance and cunning. They have also been the quarry of hunting expeditions, including one led in the 1920s by Roy Chapman Andrews of New York's American Museum of Natural History. Even today, kulans are still persecuted, though not legally. Like zebras or feral horses in the western United States, free-ranging equids are disliked by herders, because they eat grasses, as do domestic stock.

SITUATED BETWEEN SIBERIA and China, Mongolia has rotated between Chinese and Russian governance. In 1996, the Mongolian Motherland Democracy Coalition won elections, and the economy shifted from central control under Communism to one based on a free market. A large Buddhist community, many following Tibetan Lamaist principles, still prospers despite the massive destruction of most temples by Russian and Chinese forces. Since Communist rule ended, the Dalai Lama has visited the country seven times.

To my delight, he was interested in wildlife, something which became clear during a visit not to Mongolia but to Wyoming. When asked what he wanted to see by the actor Harrison Ford, who owns a ranch in Jackson Hole, the Dalai Lama answered with a single word, "moose." So, Mr. Ford drove him to Grand Teton National Park, where tourist traffic slowed their progress to a crawl. He explained what happened next.

> I'm going to show His Holiness some moose. We turned . . . where there is a big sign that says "Moose." I didn't know how to explain to him that the sign pointed to the *town* of Moose. . . . So the sign says "Moose" and sure enough, a half mile down the road, a bunch of cars had pulled over. . . . There were about three or four or five monks by the time we all got out of the cars. I don't know if he expected to see a moose because of the sign or what, but he was delighted. He stood there for half an hour watching a moose and her calf.

IT WAS OCTOBER as our plane descended from Korea into the dusty haze of Ulaanbaatar. Horses, *gers,* and belching smokestacks dotted the capital's outskirts. Autumn was in gentle splendor, as the setting sun ignited a fiery glow on mountains dappled in reds and yellows. Forests of larch, fir, and aspen mixed with open glades.

My expedition mates assembled into a small, dank room, located on a trash-filled street in the capital. With throngs of human loiterers outside, everyone assured me that safety was of no concern. I wondered, then, why two guards were needed. Identical tenement apartments stretched beyond, remnants of a communist legacy. Ultimately, our group would have 13 people, our gear, and food for weeks all stuffed into a small Mitsubishi, a Nissan, and a Land Cruiser. Since petrol stations were absent, we needed to carry the equivalent of eight 55-gallon drums of fuel, which we would use a flatbed truck. We dubbed that vehicle the "rolling bomb," due to our driver's propensity to smoke contentedly while transporting nearly 2,000 leaky liters of very flammable liquid without regard for safety or the swirling fumes.

All roads would be dirt, some just two-tracks. Our bearded leader, Rich Reading, spoke fluent Mongolian, held a PhD from Yale, and had developed wildlife projects here for ten years on behalf of the Denver Zoo. Joining us would be members of the Mongolian Academy of Sciences, including the country's premier mammal specialist, Sandvin Dulamtseren, and American biologists Brian Miller and Steve Cain.

Fifteen miles beyond the capital, the link to China broke into view—the railroad tracks to Beijing—a conduit for sharing people and goods between the two countries. The railroad also served an unintended purpose. It severed the migration of Mongolian gazelles. Also known as the white-tailed gazelle, these antelopes move across spacious grasslands like the wind. Slightly smaller than pronghorn, only males are horned. Once numbering several million, the population has been halved since the 1950s. At least 2½ million were killed for meat in China. The sweeping changes are reminiscent of the days of railroad construction and bison slaughters on the American prairies. I hoped Asia's steppes and great cultures would not follow in our shortsighted footsteps, when magnificent grasslands became fields of grain with callous disregard for wildlife migration.

Over the next week, we traveled toward the low deserts and plateaus inhabited by kulans. We crossed fertile steppes where feather grasses bent in gentle winds, where Corsac foxes searched for daytime gerbils, and where grassland marmots and Daurian pikas slipped into burrows as steppe and golden eagles hunted above. As we drifted south, the steppes gave way to broken grasslands that turned gradually into rocky outcrops. Our train of vehicles slithered through tight canyons and over rolling hills. We camped on windswept plains and along the banks of creeks where vegetation had been stripped clean by generations of livestock. At this time of the year, the weather was usually pleasant. Low temperatures would be around 20°F or 30°F. I carried only a bivouac sac and light sleeping bag.

During one lunch I offered the drivers my cache of pumpkin seeds, something they had never seen. I shared my wisdom about whether to consume the entire shell with its intact seed or to spit the shell. After a few attempts, their enthusiasm dampened. Nor did they relish atomic fireballs, those spicy cinnamon jawbreakers I enjoyed as a kid. I also introduced a Frisbee. In return, they offered chunks of uncooked fat, a delicacy from their favored fat-tailed sheep, which had been stuffed virtually whole into wrapping paper at the back of the vehicle. Our food preferences were unmistakably different.

The unexpected joys of fieldwork soon arose. We discovered an absence of oil in the front differential of a vehicle. It required nearly a day to jerry-rig repairs. Then, the flatbed with our eight 55-gallon drums of petrol disappeared, and our gas tanks naturally dwindled to near-empty. We did not know if the driver and his mate had stolen the truck, been kidnapped, or been in an accident. More likely, they were only lost or broken down.

We limped into Madakh, a rural outpost where diffident kids hid behind crumbling buildings. Three young girls dressed in turquoise-colored wraps peeked through a cracked door. With a little prodding we enticed boys, and then the shy girls, to toss a Frisbee or a football. In return, they showed us their wares, including beautiful drawings of camels, gazelles, and kulans.

Our actions achieved something else: the attention of the local police. We had needed a permit to enter this rural retreat. After describing the missing flatbed, we managed to convince them to sell us petrol from their private stash.

The mild weather broke; a dry wind blew from the west. It grew stronger and shifted south, funneling dirt and then cold bitter air from Siberia. Our tents were sealed in a fine clay and silt coating before becoming entombed in a horizontal snow. My bright yellow bivouac kept me dry, but I huddled with my warmest clothes in my flimsy sleeping bag. The drivers now slept in their vehicles. One morning, Rich surprised us.

"Let's stop for a latte."

Immediately, I snapped out of my funk, only to see team members equally perplexed. The land was lonely, sweeping infinitely in every direction. We had not seen a town for ages. My confusion showed. Quickly, Rich added "A Mongolian latte."

After more miles of driving on our way to get the Mongolian lattes, two *gers* appeared. Rich described proper etiquette. The doors of all Mongolian *gers* face south; upon entry, men must move to the west (left) and woman to the east (right). Knives were not to be pointed at people and milk not spilled. Most importantly, we were to never walk in front of an older person, nor to use our left hand to remove food from a shared plate.

As we approached, and despite the cold, two naked youngsters ran for the *ger.* Three horses in wooden saddles stared silently. A massive ram skull lay next to a decrepit Russian motorcycle—an argali, the world's largest species of wild sheep. Horns alone may weigh more than fifty pounds. An old man in traditional dress greeted us. Rich did the talking. Soon, we were inside, sitting cross-legged.

A slab of uncooked mutton hung toward the rear. We drank salty tea and feasted on goat curds and *airag*—fermented mare's milk. Vodka was poured. Ninety minutes later, we departed, warm and rather silly, but not before imparting gifts in return for the generous Mongolian lattes.

Ever so imperceptibly, the signs of humans grew sparser. Although we had passed the remains of an old military base, decaying Buddhist monasteries, and an occasional *ger,* we now entered country that was wild, appealing. Argali and ibex dotted some of the rocky breaks. Gazelles appeared but fled quickly, some in groups of twenty, others of two hundred.

We asked a man on a camel about argali in the nearby mountains. American hunters, he explained, had flown here from Ulaanbaatar. They shot from helicopters. The first argali was not large enough, so they shot another. It too was a bit runty. Finally, horns from the third satisfied the shooters.

At a different *ger,* while enjoying the usual hospitality that included warm tea, respite from the wind, and distilled mare's milk—Mongolian liquor—a man asked about our purpose. When we explained we were there to help Mongolians with studies of their wildlife, his eyes brightened. He told us how he hated wolves and hoped we would find ways to kill kulans. Even the gazelles were not welcome, as they competed with sheep and goats. I lost my sense of place, feeling briefly as if I had been transported to parts of America.

Fortunately, Namshir, a veterinarian who accompanied us, responded in Mongolian, "I like wolves, snow leopards and bears too. They are part of nature. What they eat is natural. No one should judge wild or domestic species because they all do what they do."

As we neared the border Mongolian-Chinese border, the desert floor filled with a gray vastness, rock, and rubble. Soils compacted or grew sandy. Mirages morphed into dunes and tall forests of saxaul. Stipa grasses and juniper were on hills and salty wetlands below. Tamarisk, willow, and Phragmites reeds grew along subterranean river corridors. Henderson's ground jays foraged for insects.

Camels stood out against bluish mountains. Desert gazelles grew more abundant. Known also as goitered or black-tailed gazelles, these graceful antelopes have protruding larynxes. Nearby were kulans. One group had six, another seven. Like gazelles, they, too, ran, the only sign of their former presence being puffs of dusts carried in the wind. Their flight had begun nearly a mile away. To elicit such radical escape tactics could mean only one thing—intense and repeated persecution.

"HERDS ARE SKITTISH. The animals know that people and cars mean death." So wrote George Schaller in 1998. Kulans, camels, yaks, ibex, and argali all behave similarly. Persecution and shooting from vehicles has had a less than distinguished history. When Roy Chapman Andrews visited Mongolia in the 1920s, he used primitive motor vehicles and camel caravans. His observations help explain the genesis of long-distance fear.

> They [kulans] looked very neat and well groomed in their short summer coats, and galloped easily. . . . [B]y the time they were in sight . . . we had opened fire, but they were beyond our range for accurate shooting. To me the most amazing exhibit was the endurance. . . . The stallion which we followed traveled twenty-nine miles before it gave up. The first sixteen were covered at an average speed of thirty miles an hour . . . there was never a breathing space. . . . Finally, he stood quietly and Shackleford decided to lasso him.
>
> Gazelles have amazing vitality and even when badly shot can continue to run. With a broken foreleg the animal can easily reach thirty-five miles a hour; a shattered leg slows it up considerably.

Others have also commented on the extreme skittishness of central Asia's mammals. Even on the remote plateaus of Tibet, populations of camels, chirus, gazelles, argali, yaks, and wolves have collapsed. While the decimation is caused by humans, some animals were cautious even before the slaughter began. During the 1870s, Prejevalsky noted the argali "is an exceedingly wary animal although hardly ever hunted; the Mongols finding it useless to attempt shooting them." Although wild sheep tend toward extreme vigilance anyway, others were less so before well-armed hunters arrived.

In Tibet, Hedin found docility among chirus in 1907: "antelopes were seen grazing and not shy at all, as nobody kills them there." Traveling through China's Qinghai Province forty years later, Andre Migot wrote similarly of wildlife in his book *Tibetan Marshes*: "a tremendous lot . . . which in effect is a sort of sanctuary undisturbed by man. . . . [Y]aks, wild assess, and gazelles were all quite easy to get near."

By the 1980s, this had changed. The wariness of survivors was extreme, as Schaller aptly noted the large flight distances from people. But animals have thwarted predators long before humans even arrived. In 1922, Andrews commented on such tactics. "Like desert gazelles, the 'khulons' [*sic*] seek flat plains upon which to drop their young. They are particularly careful to avoid a region of ravines or gullies which might give cover to wolves."

Most animals rely on behaviors that improve chances of their young's survival. One component involves avoiding predators, another finding food or water. Sometimes the tactics go hand-in-hand, other times not.

An individual gazelle could easily ponder, *should I belong to a big group or a small one? Dare I go to this waterhole where reeds are great and wolves might hide or lumber to another two dozen miles away?* Indeed, the tradeoffs are many, although animals may not consciously follow the same decision-making processes that we do.

Wild sheep and goat mothers typically give birth alone, among steep cliffs that wolves and foxes cannot easily navigate. In these areas, the vegetation is often sparse and nutrition limited. The tradeoff is clear: reduced access to food in exchange for offspring's survival. Other strategies are also used.

Mongolian gazelles and chiru bear fawns that immediately camouflage themselves in raw dirt or grass. They soon follow their mothers. Within weeks, huge groups appear. Like wildebeest or caribou, these Asian mothers are surrounded by youthful bleats, babies gamboling and running, jousting and butting.

Across continents, grassland parallels are striking. Pronghorn fawns become flattened blobs that blend amorphously into grass or sod. As they age, mothers group with other lactating does. Bison females do the same, mothers ultimately associating with other lactaters. Bulls differ. They remain by themselves or associated with other males, but in much smaller groups.

Both black-tailed and white-tailed gazelles form unisex herds throughout much of the year. Those of females contain adults and young, sometimes a few of which are males. Those with adult males are far smaller. Such groupings—known as sexual segregation—characterize yaks, ibex, wild sheep, and most types of large deer except moose.

Terry Bowyer championed the idea that sexual segregation is the result of coping strategies. Males and females do things very differently, sometimes appearing to an outsider as if the genders are different species. Groupings of males and females evolved due to different pressures, females and their offspring perhaps being more sensitive to predation whereas males, freed from the burden of rearing offspring, are driven more by their search for food.

In gregarious species, male groups tended to be smaller than those of females. Even if a species behavior were unknown, had individuals perished en masse, the sex ratios of the dead might offer insights about the social groupings of the living. Little did I realize how the coping strategies of the sexes and assemblages of animals would guide my thoughts as we drove south from the rural town of Madakh in search of kulan.

AS WE CROSSED sandy desert basins, the wind stiffened, propelling stinging grains at thirty miles an hour. The clouds promised snow. We passed by troughs, some used for centuries by camel caravans. After lowering buckets deep into the earth and hoisting them back up on rope made from camel hair, we had water for the next few days.

I jumped in the rear of our vehicle and closed my eyes. I thought of

Genghis Khan and Mongolian domination of central Asia. Then, I slipped further back into time. This had been a land of therapsids, a place where the dangerous cousin of *Tyrannosaurus*—the *Tarbosaurus*—pursued big prey a hundred million years earlier. This was a time of dinosaur eggs and nests, an environment of swamp and hardpan, one with herd-dwelling herbivores. I imagined domed-headed hadrosaur males chasing females and wondered whether mothers protected their young. We rounded a turn. My dreams came to life.

The stench of death filled the air. Parts of bodies appeared. As we investigated more closely, we found the spoor of foxes and wolves, the feathers of vultures and eagles. A kulan, half in and half out of mud, stared at us, its eyes hollow and picked clean. Nearby were more limbs and separated hooves. There were heads and horns. Of signs of humans, there were none.

We explored the macabre assemblage of dead camels, gazelles, and kulan. Following their tracks, I tried to cross a thin surface. It caved in, and I abruptly sank to my knees in the muddy ooze.

Despite his books and decades of travel, Dulamsteren, the stately mammalogist who accompanied us on our trek, had only once seen something similar, and never at this scale. In western Mongolia, he had counted a few dead kulans and four camels along a soft lakebed. Neither Schaller nor the accounts of other explorers reported scenes of similar carnage. Even the recent dinosaur-seeking expeditions of the American Museum of Natural History had not found modern catastrophic death sites.

Thirty-eight animals were on the surface. Twenty-five were adults—five goitered gazelles, fourteen kulans, and six camels. The cause of death was obvious—entrapment in a muddy quagmire. More than 150 years earlier, Charles Darwin had also found camels and equids together. His finds were fossilized, discovered in a part of South America where similar species had roamed before extinction.

Pretending to be a paleontologist, I asked our group to help me work through a mental exercise. I wanted to know what we might conclude about the patterns of social association had these not been dead mammals but dinosaurs instead. What, if anything, might we have deduced about their social lives based on the patterns of death we witnessed? Did they live together, in sexually segregated groups, or were they solitary animals that merely all happened to die in the same unlucky spot?

While we would never know the answer for dinosaurs, it was possible to ask the same question for extinct mammals and use living ones as a type of model. We could test it by noting patterns of life and death in the species that stared us in the face on that cold and windy day in Mongolian's Dornogovi Aimag province.

Death assemblages are exceedingly rare, but they do exist. In a Canadian bog now known as the Cursed Pond of the Big Noise, 15 moose once died. Females outnumbered males. In California's Mohave Desert, more than 30 bighorns perished in a watery trap. Groups of female elk have plummeted to death through ice, and 144 died in Yellowstone's 1988 fires. In Colorado, a group of bull elk all died from a high altitude lightning strike. More than 2,000 bison drowned in a single occurrence in the Platte River during the nineteenth century.

Moving further back in time, bison and ibex were driven over cliffs by ancient hunters in the Rocky Mountains and in the Pyrenees. Male mammoths died in groups due to entrapment in South Dakota, and Miocene rhinos disappeared under a veil of volcanic ash in Nebraska. These sorts of group deaths offer clues about the lives of extinct species. Reconstructions also hinge on the assumption that all animals died at the same time, something true for some but not all of the above.

The logic runs something like this. Most ungulates in which the sexes differ in body size also vary in appearance. Males are usually larger and carry horns or antlers, and both sexes tend to live in segregated groups. This pattern characterizes gazelles, bison, yak, and most deer. So, when groups of one gender or the other are found in a death assemblage and they are from species which are sexually dimorphic, it's likely that sexual segregation is a species trait.

Indeed, when groups of dimorphic species are tested against the frequency of one sex or another in death assemblages, the pattern holds. This aids in understanding the spatial and social aggregations of extinct species in which males and females differ in conspicuous traits.

But numerous species are not dimorphic—males and females are similar in size. Among these are most species of rhinoceroses and members of the horse family, such as *takhi,* zebras, and kulans. To these we can add peccaries and camels, including South America's guanacos and vicunas, and a group of small African forest antelopes known as duikers. None of these live in sexually segregated social groups, nor do they die in the same predictable fashion with highly skewed sex ratios. So, the ability to estimate past patterns of life in extinct species is better for those with modern analogs like gazelles or elk who live in groups of the same sex for most of the year. By contrast, the ability to predict accurately the groupings of extinct species where males and females are of the same size and lack discrete armament or cranial adornments was far poorer.

To reveal how animals cope with their environment, organize themselves as species and as individuals within populations in both space and time, present fascinating questions. They offer opportunities to pursue questions in ecology and evolution. Most notably, an understanding of the past can help guide us forward.

MY BURGEONING INTEREST in the behavior of long-extinct mammals was not shared by others in our group. Our mission focused on kulan, not bones. Patchily distributed across the southern Gobi and occurring at times in large groups, nowhere were kulans readily abundant. The good news was that they had not declined by the mid-1990s.

Rich Reading continued to focus on the possibility of a study of kulans and, because we needed more supplies, we swung back to the dusty streets of Madakh. A surprise awaited us. It was not the missing flatbed truck that carried our petrol and had disappeared weeks before. Rather, a blue van sat there with our missing fuel. We were delighted to find that the driver had not been kidnapped or injured in the type of accident that driving the "rolling bomb" might invite. It was just one more breakdown.

We pressed on. Mountains with snowy peaks swept into view. Known in English as the "Three Beauties," the Gobi Gurvansaikhan is a large desert park with tall mountains rising steeply above desert steppes. More than 2½ times the size of Yellowstone, the Three Beauties is home to a few Gobi bears and an occasional wild camel. Fitting into the wild landscape are massive dunes and saskaul forests, hedgehogs, and about one thousand *gers* that house 4,500 herders. At lower altitudes, precipitation is scant—about four inches a year. Higher, there is snow. In a gorge called the *Yolyn Am*, ice had yet to melt.

Gazelles and kulans generally avoided the areas used by herdsman. The signs of wolves were few, but as we climbed higher into the mountains, the scat of a different carnivore occurred: the snow leopard. As a predator of ibex and argali, Rich and his Mongolian collaborators noted it was human, not animal, carnivores that were decimating the wild ungulates. In the journal *Wildlife Management*, they wrote:

> [B]oth poaching and overgrazing are prevalent throughout most of these protected areas. More active management is necessary, including active antipoaching activities. . . . Sport hunting should be permitted only if argali populations are carefully managed and deemed capable of sustaining such harvests.

Although gazelles, saiga, and argali may be targeted by poachers, the species of ubiquitous disdain are wolves. They remain unprotected in wildlife reserves and are harvested from national parks. To the west in Kyrgyzstan, a helicopter-borne campaign has been underway to exterminate them. In the 15 years since dissolution of the Soviet Union, unofficial reports stated that wolves populations had dropped fifty percent. Official reports do not exist.

Snow leopards also range across central Asia. Evoking images of rugged mountains, cold, and snow, these lavishly spotted cats drop as low as three thousand feet in Gobi or as high as sixteen thousand feet in the Himalayas. From Tsu-Tai, Mongolia's southernmost glacier, the craggy Altai Mountains stretch northward to the fortified borders of Russia. Kazakhstan is to the west, China to the south. For the few seminomadic herders who graze their stock on the slopes of Tsu-Tai, the mountain is sacred. Beside their sheep, goats, and yaks are wild ibex, marmot, and wolves. In the deserts below are fleet but highly endangered saiga.

From just below the dome-shaped glacier where I have climbed, three cats have crossed the wind-crusted snow. Like tigers and jaguars, the spots of snow leopards enable individual identification. While I missed sighting them, the evidence was unmistakable—the four-toed prints of a female and her trailing kitten. A mile further, a large male had crossed the ridgeline.

To my south, the spotted cats strike a cord of fear in the herders. It is not that they are concerned about the survival of the snow leopard's mainstay—ibex—or attacks on themselves. Instead, it is the safety of their livestock. In one region, 20 percent of the resident horses and another 12 percent of the yaks had been killed.

One way herders can protect their livestock investment is to minimize predatory attacks on livestock by assuring high numbers of ibex. Another method might be to change husbandry practices. Despite the smaller size and coincident vulnerability to predation, sheep and goats are brought in at night, where they are offered more protection than horses or yak, which graze at higher elevations and where chance encounter with snow leopards are greater. To net a change in livestock husbandry is more difficult, because it would necessitate an alteration of centuries-old pastoral traditions.

Recent surveys reveal that 14 percent of herders have hunted snow leopards. To reverse the practice, a conservation program called Snow Leopard Enterprises has now been established. In exchange for reduced persecution, economic incentives are being created to develop handcrafted wool products. These are then marketed at 15 to 20 times the rate of raw fiber. More than two hundred families participated, and products are marketed locally and in Ulaanbaatar.

Throughout central Asia, all cats, big or little, confront survival on a precipice. The manul, otherwise known as the Pallas cat, is no exception. This thickly furred felid with a beautifully striped tail inhabits deserts and mountains up to sixteen thousand feet. Occurring from Afghanistan to Russia, Pallas cats are opportunist feeders. Near Lake Baikal, however, pikas make up almost ninety percent of their diet. While nothing is known about how pikas behave toward these predators, one morning I garnered information on how these cats respond to one of their chief predators—humans.

Out in the western Gobi, I left my tent before sunrise. In rubble and stone and hunkered not more than two hundred yards away was a ball of fur with a darkly ringed tail. With ears flattened and body lowered, no muscles twitched. I moved to within five yards, but still there was no further response.

The Pallas cat shares the same defensive behaviors known in other small felines. Over evolutionary time, the survival of this group of cats must have been enhanced by freezing rather than by fleeing. Whether this dynamic will change given the current harvest of Pallas cats for their prized fur is unclear, but it seems unlikely since the scale of time over which natural selection operates is long when matched against the recent spate of take by humans. The behavior of freezing and the protection offered by camouflage may both be adaptive throwbacks from the past. How effective they may prove in the future is less certain.

BEYOND WATCHING CAMELS and carnivores, the team and I had other immediate desires. Although the food we ate had lost most of its allure, cleanliness had not. We had been dirty for weeks, products of mud, snow, wind, and grime. We entered Dalanzadgad, a desert city once on the travel path of ancient Turk traders. Before looking for a shower, we headed to the local open-air market, spurred by an interest in whether carnivores were still being harvested. All we saw were the hides of goats and sheep. A good sign.

The city of twelve thousand had little fuel. Power was rationed and unavailable except from 7 PM to 10 PM. Any shower would be cold. The public sauna was closed for another week. Intent on hygiene, we opted for other public facilities, hoping they would not harbor an excess of bacteria. By the time I emerged from my cold shower, I felt clean and spiffy. Then I looked in a mirror. A wild animal had crawled from a cave.

The local teenage girls waiting their turn for showers did not know how to react to this "thing" that greeted them. They stared and giggled, noting my many weeks' worth of stubble and my dripping mop of snarled, matted hair. I heard the word *alma*, and more laughter followed.

Namshir explained that *alma* were mysterious hairy people that lived in the mountains, much like yeti or the abominable snowmen. I wondered about their existence. Rumors of large hairy primates more elusive and rarer than snow leopards have persisted for centuries.

In Kansu, Prejevalsky was once enticed to a temple to view a yeti. What he saw instead was just a poorly stuffed Asiatic black bear. In 1942, a similar story had a very different outcome. Five World War II refugees escaped from a gulag near the Arctic Circle in Siberia. Led by Pole Slavomir Rawicz, their journey by foot took them across Russia's taiga, Mongolia's northern mountains, and into the Gobi Desert before reaching India. Their trek to freedom took about a year and a half. As they descended the Himalayas and having never heard of yeti or abominable snowmen, they described an unforgettable scene:

> As we approached nearer . . . two points struck me immediately. They were enormous and they walked on their hind legs. . . . Their faces I could not see in detail, but the heads were squarish and the ears must lie close to the skull because there was no projection from the silhouette against the snow. The shoulders sloped down to a powerful chest. The arms were long and the wrists reached the level of the knees. . . . It would have been easy to have seen them waddle off at a distance and dismiss them as either bear or big ape of the orang-outang [*sic*] species. . . . There was something of both the bear and the ape about their general shape but they could not be mistaken for either.

Other reports also exist. Alan Rabinowitiz, a scientist with WCS who has spent parts of two decades in northern Myanmar, cited the report of an old hunter from about fifty years ago. A yeti had "rushed down the hillside with fangs bared and hands raised to attack him." While some reports may mistake bear footprints, which are superficially humanlike, especially when melted out in snow or dampened in mud, not all can be so easily dismissed as bears.

In December 1999, investigators from the Chinese Academy of Sciences intensified their search for a yetilike organism called "Yeren" in the forests of the Shennongjia Nature Reserve in the Hubei Province. Following on the trail of a sighting by a local hunter, the team found 16-inch footprints and reddish hair. Over the past decades, hundreds of sightings have been made from the same general region.

I asked Dulumsteren, the elder statesman of Mongolian wildlife, whether he believed *alma* existed. He replied, "in 1976 I was in southeastern Mongolia. We visited a nearby village, and I was told that a long-haired, very hairy person had sat on a woman during the night. She woke up screaming, and several people reported seeing the *alma* run off." Still, verification of a new primate, just as with Bigfoot—which turned out to be a hoax—has not occurred. Such reports are rare but widespread and tend to recur over time from the same areas. We all anxiously await DNA analyses of the red hairs.

ON THE PLANE to America, I thought of Mongolian deserts, grasslands, and mountains. In 1,800 miles there had not been a single fence. Yaks and horses, goats and sheep were all organically fed. The wild animals roamed far and wide. There had been lammergeyers and griffon vultures, kulans and lynx, gazelles and ibex. People were extraordinarily gracious, amenities few. Studies would be difficult but not impossible. Opportunities would be as wonderful and as challenging as the land itself.

Little of Mongolia's wildlife tolerates human presence. Gazelles, saiga, wild camels, and khluon flee from vehicles at a mile or more. Populations of elk, wild boar, and marmot have collapsed, most up to ninety percent in the last decade. Ibex and argali fare better, in part due to their trophy status. With stealth they can be approached more closely but only because their rugged habitats offer places to hide. Throughout the country, the price of persecution is meat slapped on the table and hides for illicit trade. Today's successful prey were either shy to begin with or they learned about knell of death cast by humans.

I recalled my last night on the steppes. As the sun dropped westward into a faraway land, I kneeled on a ridge anxious to savor the rich taste of stillness. A gentle wind fluttered. I inhaled the sweet smells of grass and the poignant dung of what I imagined was *takhi*. The night crept toward me.

A single rider came down from a remote pass and crossed the silent steppes. I remained motionless. A lone star shone in the darkening eastern sky. Snow covered the ground. Only the sounds of pounding hooves punctuated the calm. The horse crossed a ravine, and suddenly stopped. The man turned his head. Our eyes locked. He raised his arm toward the sky. I did the same.

Like a ghost, he vanished over the next ridge. A song bellowed across the empty plain. Its allure was deceptive—the cry of the nomad, the call of the steppe.

chapter 14

The Silent Cats of Patagonia

[T]he guanaco strategy of retreat. They never hide. . . . Perhaps this is not quite so dumb as most of their doings, since it might well be an ideal system against their primeval enemy, the puma, but it is still worse than useless against their present archenemy, man. Still, they have only known man for a few generations and their learning ability is practically zero.

GEORGE GAYLORD SIMPSON (1934)

IN THE WILD, cats and dogs are each first-class predators, their roles transcending more than the mere creation of fear. Apex carnivores have deep community roots. To appreciate their overriding ecological importance, one can ask a simple question: if only a single species were removed, would the effects of that loss be greater or less than those of removing another species? In essence, which change would destabilize the living landscape more, the loss of cats or dogs?

To value the connectedness between predators and nature's food webs, intricacies linking orcas with other species in the North Pacific are telling. As the most feared oceanic mammal, orcas eat other whales, including gray, blue, and bowhead, as well as an incidental moose or two. Once the whales are sufficiently reduced, orcas switch to other large marine mammals—sea lions and fur seals. Smaller morsels may subsequently appear in the diet, especially sea otters. But, once sea otters decline, the prey's favored food, sea urchins, go unchecked and explodes in population size. When this happens, the voracious urchins devour kelp forests, and the fish reliant on these habitats disappear. Like engineers, sea otters and killer whales play a significant role in structuring the marine off-shore community.

Orcas have one social trait mirrored by a few land predators. They live in pods, just as wolves live in packs. Given their nickname, the wolves of the sea, one must wonder whether group-living hunters trump solitary ones and therefore disproportionately affect ecosystem dynamics and biodiversity. Some argue the notion is pure nonsense, others that it has both merit and problems.

In reality, food webs are complex. At some times in some places, dogs can and do triumph over cats. But not all dogs, just wolves. Little is known about the ecological effects of other pack hunters such as Africa's wild dogs or Asia's dholes. As for cats, with the exception of African lions, none partake in serious group hunts.

The best evidence that dogs affect ecosystem dynamics comes from Yellowstone. Wolves cast a pall of fear, causing elk to stare in trepidation or scatter. The effects trickle down, with the consequent movement of elk relieving localized grazing pressures. Then again, at least one cat produces the same effect. The roar of a tiger in the Russian Far East also causes elk to abandon their immediate feeding grounds.

Beyond the realm of prey flight, the question remains: do cats structure ecological communities? The evidence is far from certain. Most of it suggests not even African lions control prey numbers, let alone communities. This may be because no one has yet looked for the right sort of ecological ripple or because lions share complex landscapes with leopards, wild dogs, hyenas, and cheetahs.

Far from Africa, a different picture has surfaced, one that involves a solitary cat. John Terborgh from Duke University has poked around the tropical forests of Peru, Brazil, and Venezuela for thirty years. Actually, he has done more than poke. John reported a massive top-down effect. At Lago Guri, a hundred-mile-long man-made lake in Venezuela, islands that had lost their jaguars and cougars also lost 75 percent of their vertebrates.

"Ecological meltdown" occurred, because plants were massacred by the now hyperabundant rodents who were no longer eaten, since the large cats were gone. Not only did seedlings fail to regenerate the tree canopy, but the birds and primates that depended upon the forest for food and shelter became extinct. Any semblance of ecological stability shifted as species ratcheted their way out of the system. Jaguars, in this instance, had had a stabilizing influence.

In other areas of Latin America, wild cats still prowl. Some are large, others small. Whether the roles played by ocelots, jaguarundis, margays, and *gato tigres* are similar to those of jaguars remains unknown. Of the world's 235 species of land carnivores, more than 200 have yet to receive more than a single cursory study.

IN CALIFORNIA'S VENTURA County, cougar sightings have increased, as they have throughout Washington, Colorado, and Montana. Even in Kansas, Oklahoma, and Nebraska, cougars are now returning. Known also as mountain lions or pumas, these cats are the most widespread carnivore in North and South America. Their distribution stretches from the southern tip of Argentina to the Yukon, from the baked deserts in Mexico to Alaska's rain-choked coastal forests, and from San Diego's suburbs to seventeen thousand above sea level on Peru's altiplano.

The big cat brings panic to the outskirts of LA and Boulder, where the nightly buffet is often domestic dog or kitty; it is also where attacks occur. No wonder the headline "Mountain Lions Move East, Breeding Fear on the Prairie" received a lot of attention. If fright was the desired outcome of the December 17, 2004, *Washington Post* article, it succeeded across America's midsection.

Pumas reach their largest sizes far to the south, on the cold plateaus and steppes of Patagonia. Males weigh an average of 165 pounds, females 100. The bulk of their diet there is neither pets nor native species. Instead, as in most of the world, food choice is a product of history, happenstance, and availability.

Among the prey in Patagonia are guanacos, one of South America's two remaining wild camels. Once numbering twenty million in Patagonia alone, they have dwindled to five percent of their original number. Replacements have come in different forms—millions of domestic sheep and European hares and rabbits. Closer to the Andes, other species have been introduced—wild boar and European red deer (known in North America as elk). These exotics entered windswept landscapes where guanacos and a fifty-pound ostrich-like bird, Darwin's rhea (locally called *choique*) had been the only large herbivores since the late Pleistocene. So, the puma platter currently consists of 95 percent introduced species. *Choique* and guanaco are minor players.

I BOARDED A jet in Los Angeles heading to northwestern Argentina. My goal was to glimpse the elusive *gato Andino*. The Andean mountain cat is reminiscent of Asia's high-elevation snow leopards. These smaller versions occur in Bolivia, Peru, Chile, and Argentina. Exquisitely adapted to cold and rarified air and with luxuriant fur coats, Andean mountain cats can be found at lofty heights—between eleven and sixteen thousand feet. With dark, stripy rosettes, its thick black-banded tail measures two-thirds its body length. So uncommon is this silvery gray carnivore that was not even described until 1865, and its first confirmed sighting was a mere 25 years ago. It has never been in captivity.

My first stops would be in Peru and Chile. In Argentina, I would see the airports of Buenos Aires, Catamarca, and finally La Rioja. A five-hour wait there would be followed by a six-hour bus ride to the quaint hamlet of Vinchina in the dry, subtropical Chaco. It would be another day or two before I could climb into the Andes.

The man who intended to meet me was a soft-spoken Argentinean. Handsome and bearded, with a black belt in karate and a intense keenness to learn, Andres Novaro worked for WCS. His current focus, and what I hoped to help with, would be the collection of cat feces in the high Andes. In addition to *gato Andino,* pampas cats stalked the arid steppes and rocky slopes. The overlap in altitude was unclear. Andres would collect scat from both species, use DNA to distinguish between them, and effectively map the distribution of this surrogate snow leopard to help develop a conservation plan.

Andres's WCS teammate was Susan Walker, who, while also armed with a PhD, he deemed far brighter and better looking than himself. She was the world's expert on a type of small chinchilla—mountain vizcachas. These stout rabbit-like, long-eared rodents live in year-round colonies and are believed to be the major food source of Andean mountain cats.

I was as anxious to hook up with Susan and Andres, as was my traveling partner and now-wife, Kim. Her studies of coyotes paralleled the Novaro-Walker efforts on a South American fox known locally as *culpeos.*

By the time we finally reached dusty Vinchina, two problems surfaced. The first was our gear. There was simply far too much—five duffels and two backpacks. We were ready for work at 15,000 feet and for nights down to –15°F. In addition to field supplies, we brought a computer, snowshoes, and jars of chunky peanut butter, something that could not be purchased in Argentina. We also had decent, non-rumpled clothing, since we'd deliver formal seminars later in our visit. Now, in the 108°F heat, our surfeit of gear was nothing more than a heavily weighted anchor.

More serious was the second issue. Andres never arrived. A torrent from the snow-swelled Rio Blanco River high up in the Andes had swept his truck away. Although Andres eventually emerged, his truck appeared only after a serious rescue operation. It no longer worked. In due course, we managed to secure another vehicle, but at Andres's urging, we left all gear behind except cold weather clothing. Finally, we stuffed ourselves into a single vehicle and crept slowly into the Andes.

We wound through river corridors and up and over foothills. Dinosaur footprints in 230-million-year-old mud appeared. Ahead was the powerful Rio Blanco. It separated us from the cordillera and our hunt for *gato Andino.* Searching for a place to cross in our the diesel-powered SUV,

Andres and I locked arms. We stepped into the muddy current, moving in parallel, inch by inch, hoping to avoid the drowning of a second vehicle. One hundred yards upstream, wet and full of smiles, we climbed out. Our Mitsubishi followed.

Aconcagua, the Western and Southern Hemispheres' tallest peak at 22,800 feet, came into view. High mountain lakes were still frozen, but spring was coming. Afoot in open eddies were tall crimson-colored birds with black outer wing tips, flocks of Andean flamingoes. Our immediate goal was to reach the San Guillermo Biosphere Reserve, where the wildlife was protected. A remote and vast two million acres, the reserve spanned altitudes from ten thousand to nineteen thousand feet. It harbors ancient Inca burial sites, including Cerro las Tórtolas, above eighteen thousand feet. Access was limited because of bridges that had washed out. Visitors were few. In 2001, there were 15; in 2003, none other than us.

We entered the Valle de Vicunas—an open valley at ten thousand feet with rolling plateaus, colored rock, and steep hills. Snow-covered volcanoes dotted the horizons where Chile loomed just below. Vicunas emerged, each in tight family units dotting the desolate plain. Once we stopped to look, the cream-colored, long-necked bodies blended in fluid motion. Yellowish aprons bobbed as these wild camels fled, a legacy of past persecution.

Like all camelids, both sexes lack head ornaments. Males possess dangerous canines that can puncture the legs of rivals during serious fights. About the size of deer and restricted to high elevations, we counted almost 1,200. Pumas also were present, several having been sighted along water courses. With no tree cover, the big cats were visible. Importantly, they consumed native prey in the protected reserve, unlike their primary foods elsewhere in Argentina.

Over the next few days, we searched for *gato Andino.* Despite thin air and a relentless sun, we checked steppes, steep hills, rocky slopes, and precipitous cliffs, all for turds with fur that had passed through the digestive tracts of unknown cats. We ducked into caves, twisted our bodies in crevasses, and slid across fields of talus and scree.

With about one hundred scat samples in hand, Andres was satisfied. We dropped out of the northwestern highlands and headed toward Susan and Andres's home in Junín de los Andes, a small picturesque town in northern Patagonia. We had not seen an Andean mountain cat, but there had been vicunas and the whistles of mountain vizcachas. We watched in awe as Andean condors floated on air above jagged peaks and marched across soils perforated by the fossorial diggings of tuco-tucos. We watched a cougar salivate during a midday sojourn across the treeless altiplano, no doubt in search of vicuna. Flocks of chick-less *choiques* fled. Four days and countless bus rides later, we arrived with our loads of gear in Patagonia.

Raw, beautiful, and the size of California and Montana together, Patagonia's population density is amongst the lowest on the planet. Badlands and steppes mix with deeply incised canyons, some sitting below glaciers. Penguins and fur seals, sea lions and killer whales use the wild coast. Across this vastness we hoped to understand more about predator-prey systems.

At Parque Nacional de Lanín, arid steppe grades into southern beech forests and bamboo. It is also where a stoutly built deer, the huemal, occurs near treeline. Further south is a smaller deer with beautiful reddish-brown coats and small spiked antlers. The coyote-sized pudu is restricted to low mountains where they fight a continuing battle. It is not so much with their ancestral predators, the puma, but one compounded by the dual pressures of hunting by humans and chasing by dogs. Competition for food with livestock has probably not helped.

Sheep estancias rule, as they have for more than a century. As a result, grazing pressure in Patagonia has been intense, forcing guanacos to abandon areas overgrazed by sheep. Since guanacos and sheep overlap in diet, the camels are about as welcome as cougars.

Sitting around a campfire with gauchos one night and riding horses with them the next day, we asked about wildlife. To our surprise, they liked guanacos, offering that the beige animals added wildness to the spartan landscapes. A few were always good. Some Argentineans even marketed their wool. Only a few decades earlier, skins were sold overseas; 450,000 had been exported during a seven-year period in the 1970s.

The next day, we discovered for ourselves the popularity of native and introduced species. At a hacienda, the antlers of two red deer were proudly displayed above the fireplace along with the tusks of a boar. In the barn, 25 *culpeo* skins and four puma hides hung from nails. This harvest was not one of abject hatred. Wildlife exports brought $100 million annually to the Argentinean economy. If each gaucho were to capture and skin ten to twenty *culpeos* a year, it would account for five to ten percent of their annual salary. Unless it paid to retain wildlife locally, conservation incentives were not likely to emerge.

The Valdez Peninsula is one area where wildlife is at a premium. Tourists visit to see elephant seals, sea lions, and fur seals, penguins and killer whales. There are monogamous maras—sleek rodents the size of large porcupines that pronk like ungulates—and hairy armadillos. There are bed and breakfasts, boats and beaches. The food is excellent, the wine inexpensive, and coastal ambiance enchanting. It is an area of sheep estancias, few guanacos and fewer cougars. It is not the native species on land that attract; only the wild and protected seascape and the marine species it supports.

ELSEWHERE IN ARGENTINA a desire to harvest introduced game remains strong. These exotic species offer unique opportunities to understand predator-prey behavior. The introduced European red deer has had no recent evolutionary history with pumas. Even though red deer in eastern Asia must face tigers and those in North American are exposed to cougars, those originating from Europe and subsequently transplanted to Argentina must for all practical purposes have been predator-naive. If North American elk and moose had both lost their fear of wolves and bears, I had no trouble believing red deer would know nothing of pumas.

Werner Fluek, a German biologist now living in the Andes foothills, witnessed interactions between pumas and red deer, reporting that red deer behaved the way they were supposed to. They became vigilant and fled. Fluek's observations depict Argentinean red deer today. No one knows how the first red deer fared when they experienced their first pumas. Perhaps they died just as easily as the first naive elk in the Tetons did when encountering their first wolves. What is clear is that red deer now respond. How and the period over which those responses developed remains unknown. Perhaps red deer are generalists, tuned merely to the universal body plan of four-legged squat predators, or perhaps they have learned about pumas, per se.

Also unclear in this part of Patagonia are the ecological effects of introduced deer. Although the silent glance of a cougar or jaguar might cause prey to shift about, population control of exotic species is not likely to occur when predators are heavily harvested by humans. The issue, while highly relevant within Patagonia, reaches far beyond its borders.

At least three species of deer were introduced into the Southern Hemisphere. Fallow deer have been in southern Africa since at least 1869. Red deer and moose were planted in New Zealand in the 1920s. The moose died out but not the red deer. Their once-uncontrolled populations caused massive erosion and damaged native forests. A different issue concerns the deliberate introduction of native species much closer to their normal haunts and what this reveals about food webs, behavior, and the possible primacy of predators.

SURROUNDED BY THE boreal waters of Canada's Maritime Provinces lies the island province of Newfoundland. Its vegetation is suffering. Spruce budworm has plagued balsam fir for decades, saplings are mostly absent, and the woods are a skeleton of their former robustness. At the opposite end of North America, out on Kalgin Island in Alaska's Cook Inlet, the story is much the same, a decaying forest. Far to the south in Colorado, riparian zones that were once thick with willow and cottonwood are likely no more. It is not global warming that is turning these once fertile forests decadent. Rather, it is an introduced browser—moose.

Why the shy giants had not reached these areas in the past is of interest, as is much about a species' distribution. Whatever the explanation for their historic absence, the fact is that they did not colonize these areas on their own. Neither did large predators.

The introduction onto Newfoundland began in 1904 with four moose. Today there are more than 125,000. The total numbers are far fewer at other sites, but population increases help frame the logic underpinning apex carnivores and food webs.

In the absence of top-down control by wolves and bears, or the jaguars and cougars of Lago Guri in Venezuela, some large mammals have the capacity for rapid population growth. Moose are but one, white-tailed deer another. As Thomas Malthus wrote in 1798, no population can increase infinitely. A food ceiling will inevitably limit population size, as we know from studies spanning flour beetles in the laboratory to elephants in the wild. If it is not food or predators, then disease or drought, fire, or a brutal winter; other factors will play a role. When the habitat can no longer support a population, the population will affect the habitat. This is the point at which the lack of predation coupled with high herbivore densities impacts biological diversity.

Terra Nova National Park sits on the northeast edge of Newfoundland. Moose densities are high and mountain ash, birch and balsam fir have been brutalized almost everywhere. The exceptions are moose-proof exclosures, with fences nearly ten feet tall. On Kalgin Island, moose browse on roots. They search beaches for seaweed, kneeling to ingest the rich supplement to an otherwise meager and overbrowsed woody diet. As Terry Bowyer pointed out, the teeth of four-year olds were so chiseled the animals appeared as if they were ten-year olds. The bottom line was that voracious browsers eat what they can to stay alive. As preferred foods become less available, the diet may expand until there is little left for the starving beasts.

These are the likely break points where effects cascade from one ecological level to the next. Even on Isle Royale, where wolves prey on moose, deciduous forests have given way to coniferous ones. In Sweden, high moose densities have changed the face of forest leaf litter and limited the diversity of ground-dwelling beetles. And, prior to wolf and grizzly bear expansion into the Tetons, high densities of moose not only altered the physical shape of the willow canopy but, because of their extensive browsing, excised a large toll on migratory birds. Calliope hummingbirds, willow flycatchers, gray catbirds, yellow warblers, black-headed grosbeaks, and several species of sparrow were all reduced or absent from areas with high moose densities. Just outside the park where hunters had replaced grizzly bears and wolves—removing 10,800 moose during a twenty-year period—the vegetation was more vigorous. Birds nested at higher densities.

What all this means is that carnivores can and do play important ecosystem roles that ultimately affect levels of biodiversity. It does not mean that predators everywhere are always controlling ecosystem processes or are the most important species. Runaway ungulate populations, whether introduced onto islands or through natural expansion, have the capacity to reach elevated densities even in the presence of large carnivores. Less appreciated is how herbivore populations have the capacity to drive carnivores to extinction.

Anticosti Island, a Yellowstone-sized chunk of maritime forest in Canada's Gulf of St. Lawrence, was first described in 1542. On it were just two vegetation-feeding mammals, black bears, which are omnivores, and mice. Exactly three hundred years later, a fellow named M'Ewan wrote, "I have never seen any place where the wild berries, of various kinds, are so plentiful." In 1896, two hundred white-tailed deer were introduced. By the 1930s, the population numbered more than fifty thousand. Steeve Cote, a professor at the Université Laval in Quebec showed that the extreme browsing reduced berries and other vegetation favored by bears to mere nubs. He also suggested the remaining plant fodder was insufficient for fat accumulation and that, "In the absence of other evidence to explain the decline of bears . . . white-tailed deer likely extirpated black bears."

Where concern exists about overly overabundant species, a common argument is that hunting substitutes for the loss of predators. In a 2005 *New York Times* piece, titled "Who's Afraid of a Few White Tailed Deer?" Anthony Licata wrote, "we [humans] are as 'natural' a predator as wolves or mountain lions. Sport hunting is safe, effective, and ecologically sound." But are the ecological services of cougars or wolves really so easily replaced?

At one level perhaps, but not at not others. Human hunting can check ungulate densities, thereby facilitating biological diversity. However, as a mimic of natural predators, harvest by humans differs in critical and nontrivial ways.

Carnivores hunt year-round, not seasonally. Scavengers access carrion throughout all seasons when natural predators exist. Carnivores kill more young by an order of magnitude than do hunters. Carnivores also kill other carnivores and, because of the threat of predation, they shape the movements of smaller predators. The infrastructure to support human hunting also differs. We humans tend to rely on horses, offroad vehicles, trucks, and often elaborate field camps to get into the back country. We erode soil and change water regimes, trample plants and shoot unprotected varmints like porcupines, skunks, and coyotes in the process of being out in the field. Hunters, however, are far outnumbered by tourists, mountain bikers, and backpackers who do not trod lightly on the land either. The impacts of people are many.

LIKE MOST FELIDS, the ecological role of Patagonia's sole big cat remains little known. Although guanacos or vicunas are the major prey in a few large reserves, beyond these porous borders domestic and introduced animals become the main cuisine. To derive an understanding of how or if predation structures landscapes along lonely coastlines or in the high Andes will be difficult.

While carnivores that are large and dangerous receive a fair amount of attention and persist locally in a few areas of the world, less than 15 percent of this flesh-consuming group numbering about 235 species have been the subject of much study. Nonetheless, aggressive conservation efforts are underway. They vary as much among countries as among continents. Australians are trying to put a tight reign on the march of aliens and still maintain or restore native species. In Mongolia, problems associated with invading and feral carnivores are not the issue. Persecution is, with predation by carnivores now being replaced by heavy and illegal harvests by humans. Whether predators or humans have differing impacts on prey is less of a pressing conservation concern than are finding ways to maintain snow leopards, smaller native cats, and wolves alongside the species they consume. Relative to the overwhelming human milieu, the silent stalk or fiery stare of a cat has had minor impacts on the behavior and distribution of prey, even in remote and unpopulated areas of central Asia and Patagonia.

part v
Making the Beast More Savage, or Less?

I am now convinced that wild animals have not always behaved as wild animals behave today. This is because our belief that animals must be tamed in order to show a lack of fear of humans is wrong, for there was a time when all wild animals behaved like tame ones. Long ago we humans made the savage beast—and ever since we have been attempting to unmake it.

TIM FLANNERY, *The Future Eaters* (1995)

chapter 15

A Credibility Conundrum

I am tired of this thing called science. What do we care about stuffed snakes, alligators, and all such things?

SIMON CAMERON (1861)

MANY FAIL TO make the link between the study of natural history and the study of science. Pennsylvania's distinguished senator, Simon Cameron, was no exception, despite a rich tradition of inquisitive pursuit by one of Philadelphia's most notable citizens, Benjamin Franklin. Cameron's frustration boiled over in 1861 in response to the congressional budgetary request of Joseph Henry, the Smithsonian Institution's first director, who wanted "to make science more respectable at home."

Even today, the rancor between science and application has grown only firmer, whether the issue is global warming, stem cells, or ecological restoration. When knowledge becomes available, the public demands it, along with accountability, a force that causes governmental practices to change. The more obvious cases involve industry's loud rebellion against regulatory actions once the deadly effects of using tobacco and agricultural pesticides became known.

While science has cast aside visions of demons and demigods, nearly half of Americans today still believe in some form of broadly defined ghost, an indictment of the discipline's nonefficacy. Misunderstanding, coupled with a declining interest in science, looms so large that one wonders how the public learns that the world is round, gravity is real, or humans are genetically one more form of a great ape.

The acceptance of cold, hard facts is far easier when they have no bearing on our culture, our beliefs, and our economies. Science is controversial, and some people will never believe a fact is a fact unless they witness it. Stubborn determination has both its merits and detriments.

No one really believes the world is flat, yet numerous people fear carnivores, because the potential threat to themselves of their children is real, even if countless more humans are killed by lightning and bees. Biophobic responses to potentially lethal species are of clear adaptive value, although today such trepidation may appear illogical or unrealistic. In our past, the basis of our adverse reactions was understandable. An unusual set of information involving lethal interactions between people and large carnivores exists for Scandinavia.

During the past 250 years, 27 people have been killed by brown bears, 17 by wolves, and none by either lynx or wolverines. Based on recent interviews, women were more afraid than men, a behavior that coincided with the greater risk of death to woman. During attacks by bears or wolves, at least 60 percent of the females died, but the death rate for males averaged about 35 percent. People from rural areas living without carnivores were more likely to be fearful than those living under conditions with one of the large carnivores, an indication that people who coexist with carnivores are less biophobic.

Despite a basis for fear and the logical call for precautions, a serious challenge remains in the United States. A base of knowledge has shifted from fact to perception. Nowhere is the mix as volatile or the views more intransigent than when it comes to carnivore reintroduction. Dirk Kempthorne, the former governor of Idaho, exemplified the juggernaut. Grizzlies had been extinct in Idaho for more than fifty years when scientists had documented a broad swath of available habitat that could support hundreds of bears. Public comments favored their return by nearly a fifty-to-one ratio. But the governor stood firm with his argument that he did not want the "bloodthirsty" carnivores back in his state.

Despite elected officials with no interest in conservation, agencies have missions that serve the public interest. Some even tout the value of science until the very same data they seek to obtain threatens their mission. Has science helped guide policy with respect to predators in the American West, or is science just another beast for a public already paralyzed by polarizing rhetoric?

HOW PREY RESPOND in systems where predators have been reintroduced is an issue that has received precious little attention. Luke Hunter opted to tackle the issue, moving from his native Australia to Africa in the 1990s. He did so because lions and cheetahs were about to be reintroduced into South Africa's Phinda Reserve, a brush ecosystem already populated with zebras, wildebeest, and nyala. Between 1992 and 1994, 13 lions and 15 cheetahs were returned. Luke concentrated on both predator and prey, finding that after just five months, vigilance in impala and wildebeest spiked two hundred percent. Cheetahs and lions worked harder for their next meals.

Luke, now a WCS biologist promoting carnivore conservation, brought forth an area of highly relevant research. By concentrating on both prey and predator, he gathered information about the perceived destruction of valued prey. Although wildebeest declined in the small reserve, it was in part due to a lack of appropriate refuge areas and habitat diversity. In addition to enhancing what is known about predation, Luke highlighted what is not—the rate at which prey respond, learn, and persist.

Information gaps impede the construction of bridges between science and the public. The consequences are alarming for conservation, but the gaps themselves are only part of the problem. The other is social. A smoldering hatred of carnivores in some bucolic areas of the American West illustrates why science itself is grossly insufficient as a force in conservation.

Interpretations of results often vary. This makes sense. Sample sizes may be constrained, and analyses may differ. But when the public does not understand science even at the coarsest levels, the challenges intensify. Often the public does not want to hear about science, because they believe it is not credible. No one denies that dialogue and discussion, disagreement and free expression are among the core values of a democratic society. Yet, the use of science should reign as a heuristic tool to guide knowledge, to separate fact from fiction, and to inform conservation.

At a local level, the story becomes sordid, unfolding in remarkably similar ways across time and space. Recalcitrant employees of public agencies gleefully turn science into personal agendas. Todd Wilkinson's book, *Science Under Siege: The Politicians' War on Nature and Truth*, documents many cases where science has been deliberately devalued, sometimes by hostile administrations, other times by the unscrupulous with a disdain for facts. In my case, it would be the declining moose of the Greater Yellowstone Ecosystem and the role of large carnivores.

In 1994 Joe Bohne, the well-respected field biologist with the Wyoming Game and Fish Department and an enthusiastic sportsman, argued for approval of my research permit to study moose. Joe did so because of an honest interest in knowledge. Nine years later, I continued meeting with natural resource managers in Joe's own agency; the mutual goal was to model populations so that harvest quotas could be established.

In 2003, a local game biologist with a strong dislike of carnivores decided to use data on the number of moose calves killed by bears and wolves in Alaska to develop a population model with which to set human harvest levels for Wyoming moose. The logic was that if the number of moose that were to be shot by sportsmen was lowered because of predators, the level of animosity toward wolves and bears would increase.

Joe pointed out that the Alaskan mortality figures did not reflect my findings from Wyoming. I went further, indicating that most mortality in my radio-collared moose was not due to bears or wolves. Empirical results were not what the local biologist had in mind. He argued that cause of death was moot and that an overestimate in juvenile mortality would allow quotas to be set more conservatively.

"Don't you think that is totally disingenuous? Modeling with inappropriate data is simply wrong. The issue isn't about management or quotas, it should be about using the best science first, wouldn't you agree?" I extended my comment, suggesting that carnivores should not be the pulpit upon which to base management decisions, especially if "blaming" them was not the cause of low calf recruitment.

Little did the public know how lucky they were to have Joe Bohne in their corner. He pitted himself against local authority; his deliberate stubbornness kept groundless assumptions from plaguing the moose harvest model. After Joe was transferred, the assumptions used in the exercises became murky.

The anti-carnivore agenda continued to play out and was captured in a 2005 guest editorial by Mason Tibbs, a sportsman and cowboy, in Jackson Hole's local newspaper.

> Am I the only one to see the correlation between wolf reintroduction times and the decline of moose populations? When Mr. Berger's findings were released, I visited with a friend of mine who works for the Game & Fish about them. . . . I know from personal experiences how hard it can be to determine the cause of death of an animal when the carcass is a week or more old. If it is dried up and hard like a rawhide, the only way to determine if it is a wolf or grizzly kill, is to take the carcass back with you and soak it in water. . . . It is my understanding that Mr. Berger did not do this with these types of carcasses, instead merely chalked them up as death unknown.
>
> With regard to the low reproductive rates, any animal that is forced to run for its life day after day, and stressed to the point of exhaustion will

> surely slough [abort] its calf. This is common knowledge among ranchers, however it appears Mr. Berger did not consider it.

MR. TIBBS WAS not the least bit off by initiating a dialogue, nor was he incorrect in suggesting that data be examined from different perspectives. He erred by not pursuing facts, many of which were available in annual reports or publications. Had he dug more deeply, he would have discovered that more than 85 percent of the carcasses that I examined were well within a week of death—most within two to three days—and that less than 10 percent of the total deaths were assigned to "unknown" causes. He would have discovered that bears were inactive, tucked quietly away in dens, when the majority of the winter deaths occurred.

THE FESTERING DISTRUST of science is a conundrum faced by all. Is science to be blindly embraced, fiercely debated, or simply ignored? Who funds it? Who performs it? Have conflicts of interests been explicated?

The cultural divide between purveyors of predator hate and those hopeful for results had exceeded any reasonable bound. It did not stop there, or in circles beyond the realm of moose or even Wyoming. Among the deliberate ways in which public perception can be manipulated is the distortion of fact in putatively scientific manuscripts. Once unpublished results of scientific study pass favorably through the peer-review process and make it into credible journals, the public is less willing to call results in question. Thus, one could imagine my disbelief when a coauthor of mine with a visceral dislike of predators inserted a sentence into a manuscript intended for publication. In the article, titled "Health Assessment of Shiras Moose Immobilized with Thiafentanil," one line stood out: "Five other moose died >4 wk post capture with no apparent cause of death determined because of scavenging of the carcasses but with evidence of predation by wolves, mountain lions, or grizzly bears." Its purpose was to incite, but the sentence was nonsensical. Without determination of a cause of death, how was it possible to conclude *with evidence of predation by wolves, mountain lions, or grizzly bears?* Of course not. Since I was to be one of six authors, I insisted the statement be removed.

Terry Kreeger, the paper's senior author and a veterinarian with the Wyoming Game and Fish Department, immediately agreed and removed the sentence. Nevertheless, it was just this sort of clever ruse falling under a banner of scientific pretext that fuels one more aspect of the continuing vitriolic attitude toward predators.

Wyoming's distaste for wolves and grizzly bears was not uniform. Its game and fish administrators in Cheyenne, along with many of their field biologists, had a genuine interest in facts. The department's director, Terry Cleveland, and its governor-appointed wildlife board invited me for a hearing in Casper. In my lecture, I described data, methods, and findings. The episode was summarized on Wyoming public radio, television, and in local papers. Opinions were mixed. People believed what they wanted. Some felt the moose collapse was because of predation, even though populations in other parts of the state were also declining and wolves and grizzly bears were absent from those. Hostile e-mails told me to quit fabricating data or to leave the state. Some of these messages were also sent to the governor and Wyoming's congressional delegates. Others, at least, thought I had been careful in suggesting malnourishment or disease as likely causes. The debate is far from reaching resolution, although data now suggest that predation plays a minor role.

Management steeped in science is one thing, public perception another. The odium for predators continues to inspire an atmosphere of anti-education, coupled with a firm distrust of scientists. Claims, such as wolves caused the collapse of bighorn sheep by chasing them just for fun or bears killed all the moose, are anything but scientific. But without effective communication to firmly anchor an appreciation for science in the public arena, blind assertions take the place of fact and fuel traditions of fear.

Mahatma Gandhi once said, "The greatness of a nation and its moral progress can be judged by the way its animals are treated." In the contentious Yellowstone ecosystem, data possession has not equaled rational discussion. Elected government officials and administrators there and well beyond often operate deliberately in dataless vacuums, choosing a safety net of ignorance in decisions that can be far-reaching. In the end, the biggest losers are the public. They deserve better; they deserve information. Science is not the beast. Misinformation is.

chapter 16

Different Sides of the Darwinian Divide

> We're certainly a dominant species . . . we're a species that when you add us, the diversity collapses. We can change everything . . . and destroy everything.
>
> MICHAEL SOULE' (2002)

WITH NEARLY SEVEN billion people on the planet, we have affected virtually all life, human and otherwise. Some ecosystems have fallen apart. Remote ones remain filled with biological treasures. When populations were smaller in our distant past, were our impacts less? Or, did we drive the blitzkrieg of fellow mammals only one hundred centuries ago? More to the point, did prey naiveté contribute to the success of North America's first hunting peoples?

First, concerning extinctions themselves, the evidence is mixed. A human role is clearly based on direct evidence in places like Madagascar and New Zealand, inferentially strong for Australia. In North America, answers are less obvious. Effects come in many forms, including fire and ecosystem change, but the debate encompassing the impact of North America's first peoples has concentrated on excessive hunting. While hard evidence is always the litmus test, it is of interest to ask whether it is plausible that colonizing humans were the slayers of the bulk of North America's megafauna.

For some species, the idea can be rejected. Habitat, not humans, enabled musk ox and caribou to persist in the far north. The native horses and saiga of Alaska and the Yukon also would not have persisted much longer regardless of human presence, since they require broad bands of dry steppe to survive, swaths lost in the north and unavailable to the south due to blockage by glaciers.

Elsewhere in North America the jury is still out, but evidence for a bipedal effect mounts. Support comes in three general forms: the undeniable hunting of mammoths and mastodons as noted by arrowheads found at fossilized carcasses, the swift disappearance of many other large species just after the arrival of the first Americans, and the subsequent transcontinental expansion of humans. In addition, the remains of extinct species mixed in ancient fire pits in the Caribbean are clear proof of a human role, although undeniably these are from island faunas, which are always more prone to extinction in the first place.

As for the premise that the naive succumbed en masse to past human hunters, speculation is plentiful but facts are few. We cannot recreate the behavior of extinct prey during their first or subsequent contact with colonizing humans. The most frequent claim that prey savvy facilitated survival throughout the late Pleistocene is echoed by a comment of the famed Scandinavian paleontologist Bjorn Kurten: "most of the European invaders of North America, the moose, wapiti, caribou, musk ox, grizzly bears, and so on were able to maintain themselves, perhaps because of their long previous conditioning to man."

At face value, Kurten's suggestion has appeal. If correct, then the converse must also apply. Species originating in North America, which will be anything but savvy to humans, should have succumbed upon first contact. Inference is all that is available to examine this idea.

Some contemporary species whose families or distant histories are of North American origin did go extinct. But not all. Some currently survive beyond North America, including spectacled bears (in the Andes), dholes (in southern Asia), equids (such as zebras in Africa and kulans and kiangs in central Asia), tapirs (in southeast Asia and Latin America), camels (in pockets of Asia and Latin America), and saiga (in central Asia, though their evolutionary origin was not North America). The idea that it was behavioral enlightenment that promoted their success is highly confounded. Ungulates and carnivores differ in so many important respects that the strength of any single overarching explanation for their persistence is limited.

Reliance on living analogs is not especially helpful here, either, since the degree of current disturbance by humans is far from trivial. In their non–North American environments, hunting and habitat change, threat of persecution, and alterations of entire ecological communities are all likely to have shaped current behavior of surviving species at least as much as events of the distant past. These and other factors render it difficult to know much about the role of prior naiveté in past extinctions.

There is yet another complication. Not all hoofed mammals endemic to North America suffered extinction. Today's pronghorn survived the blitzkrieg; related genera did not. The human overkill hypothesis predicts the swift loss of naive large mammals, though it is unclear whether every neophyte species must succumb for the idea to be tenable. What, instead, if only 90 percent or, say, 75 percent of the species or genera had been extirpated? Would human overkill be falsified? By chance alone, a few non–human savvy species might just persist.

As for elk, moose, and the other Eurasian voyagers that crossed into North America, we now know that some learn rapidly. The mélange of effective defense is reinvigorated quickly once native predators return. As to whether colonizing Neolithic humans should be lumped with native, historical four-legged predators is a different issue. There is nothing that can be known with certainty about how prey detected past predators. Some could have viewed humans as predators and through experience quickly learned about them. Others may have been less astute and died out as a consequence.

SHIFTING FROM THE past into today's tangled conservation arena, the implications of melding fear and naiveté broaden. Apprehension of death guides behavior in similar ways whether on the human or animal side of the Darwinian divide. Given our logical human predilection for our own safety and the risks of coexistence when living with perilous species, we ought to know whether we have made animals more dangerous, not less. Has their persecution increased our safety?

As early as 1914, Theodore Roosevelt believed this was the case, from his experiences hunting in East Africa: "When lions are much hunted it is doubtless true that they grow so wary of man that only the dire . . could make them think of preying on him; but where they are less molested, their natural ferocity and boldness . . . will take to man-eating." Thirty years before that, Roosevelt had made similar suggestions for grizzly bears on the American prairies.

An intuitive response as to whether we have made "beasts" more wary is "yes." Where tigers or bears are harassed or shot, the fearless are the first to be eliminated. As logic dictates, warier survivors would pass on more knowledge and genes, abetting the next generation in its avoidance of us. This sort of hypothesized adaptive response is precisely why, some argue, brown bears on the eastern side of the Atlantic are less ferocious than their more aggressive North American cousins. Europeans removed the bold long ago; perhaps the meek are all who remains.

A different case, though with similar rationale, emanates from California. Persecuted for as long as anyone can remember, cougars have been legally protected from hunting for about two decades. They are now more visible. Since 1990, several joggers and a mountain biker have been killed. In contrast, less than twenty people died from cougar attacks in the entire United States. during the prior hundred years. It seems likely that a loss of persecution has prompted the large cats to become less shy. Habituation to people has led to a change in behavior and a change in food habits.

Carnivores are quick learners. In Yellowstone and the Tetons, neither wolves nor coyotes have been hunted, and they are relatively easy to capture for the placement of radio collars. However, all it takes for wolves to seek the safety of forest cover is a single chase by a helicopter. Coyotes show parallel abilities for one-time learning. Individuals from non-trapped populations know nothing about snares yet, but after being snagged once, the chances for recapture drop to levels similar to those in persecuted populations. Even lizards, such as iguanas on Galapagos that live without fear of humans, develop physiological responses indicative of the behavior of avoidance after one-time capture or harassment. Quick learners have an obvious advantage.

The transition from fear to tolerance emerges in similar ways across continents. In Africa, spotted hyenas have experienced periods of boldness that resulted in danger to humans. In some areas, hyenas first became accustomed to feeding on the trash spread across common refuse areas. As fear of humans diminished, interactions increased. During the mid-1950s in southern Malawi, 27 villagers were killed.

At Lake Tahoe along the Nevada-California border, the consequences have been less extreme, but parallels with hyenas are strong. Coyotes are afforded no special conservation status, but in developed areas with casinos and expensive homes, the shooting of guns is prohibited to safeguard humans. Without persecution, coyotes grew bolder throughout the 1990s. Tourists and well-intentioned residents began offering food, first scraps but later food left in bowls. Some coyotes were even offered food by hand. One person was bit and several children stalked. Finally, the respective management agencies stepped and reversed the habituation.

A more potent risk to human safety is posed by the grizzly bear's smaller cousin, the black bear. Near the Liard River in northern British Columbia, a 37-year-old mother traveling from Texas to Alaska in 1997 was killed by a black bear, as was the man trying to save her. Two others were severely mauled. An employee at the nearby lodge said, "I think it's pretty much one of the most brutal ways to die, having your entire body ripped apart."

Nonpersecuted bears are more likely to attempt predation on humans, although exceptions always occur, and people are killed by habituated ones, too. Bears often associate food with humans, but they rarely treat humans as food, per se. The individual history of the Liard River bear was never known, nor was it clear if the deaths were due to attempted predation or defensive maneuvers.

The population of black bears inhabiting Lake Tahoe's pine forests offers a differing and more in-depth view of the changing relationship between people and carnivores. They do not seem to fit any model, a difference that underscores the difficulty of neatly packaged explanations. Those on the Nevada side have never been hunted. Until recently, they remained shy, living primarily in the mountainous back country of the Sierras where encounters with humans were few.

Their changing patterns of land use have been tracked by a certain ex-Kansan. Equally at home behind a tractor, a computer, or a hunting rifle, and armed with a doctorate in biology, Jon Beckmann of WCS has rooted himself in bear biology. Jon believes bears are inherently good, though he cannot say the same of their recent trash-raiding habits. He argues that we should manage people, not bears.

Jon and his colleagues at the Nevada Division of Wildlife wanted to reverse the current practice of bears diving into dumpsters for garbage. Their goal had been to turn nonhunted bruins back into wilderness foragers, something they had been up until 1989.

Ten years later, not only had most of the back-country dwellers discovered trash, but many became so comfortable around residential dwellings that nearly a dozen denned below the porches of the multimillion dollar mansions lining the emerald lake's shores. The bears were habituating and presenting clear dangers to humans. To counter these emergent behaviors, Jon and his partners experimented, using nonlethal deterrents—dogs, capsicum, and rubber bullets—to drive bears away. None were effective. The harassed bears regularly returned. The fear factor, if any, was not activated. What did work was a campaign rooted in public education and outreach, followed by county-wide ordinances mandating bear-proof dumpsters.

The result's of Jon's biological study were interesting for what they said about the adaptable nature of bears and how best to conserve those in the Lake Tahoe Basin. But they had little to do with a more general understanding of whether mothers passed along information to their cubs, in particular about avoiding humans or finding food.

Beckmann's interest in the transfer of information between generations took him and his coauthor Stuart Breck a step further. They asked whether female cubs adopted the "bad" behaviors of their mothers. Surprisingly, the answer was no, even though bears learn quickly. A mother's trash-feeding habit was not necessarily adopted, nor did female cubs of back-country foragers remain that way. Young females were not automatic replicates. Differences between mother and daughter could arise for reasons associated with dangerous males at dumps, an inadequate distribution of natural food, or such low social status that chances to obtain trash were limited.

While some Tahoe bears attain weights in excess of 750 pounds, they may or may not build up fear of people. Bears and others members of the carnivore family do typically develop a healthy respect for native predators. Simply because an animal kills others so it can eat does not somehow render it immune from death by other predators. For instance, with the reduction of grizzly bears and wolves from some 98 percent of the contiguous States, only a few larger carnivores persist. The public at large generally assumes that black bears and cougars are not vulnerable to death by species other than humans. This would be incorrect. Even the bad and the dangerous of the most feared species have trepidations as they roam about as individuals. Their movements and activity patterns are modified. Certain areas are avoided, and habitats that offer greater safety are preferentially used. Black bears are not frequently in areas where trees are lacking. Why? The risk of fatality increases.

Grizzly bears are known to kill and sometimes consume black bears. Wolves defend their dens against grizzlies, even attacking them, but when opportunities arise grizzlies are known to kill wolves. Spotted hyenas do not like lions and vice versa. Indeed, the presence of some species alters the behavior of others, whether carnivores are writ big or small.

So what might render a large male grizzly cautious? One possibility is a larger predator. While this cannot obviously happen in North America, in the Russian Far East it does; tigers kill bears. A second and more likely possibility is an encounter with another male grizzly. Sooner or later, the most dominant or the most aggressive animal in a population will lose the fight. There will always be a subtle line between accessing a limited resource (females and food are two examples) and the tradeoffs associated with accessing such prizes. The risks of injury or death offer powerful incentives to change one's behavior.

Mammals other than carnivores also pose dangers to humans, especially species of large size and testy disposition. For moose, some information is available on how behavior changes. From the Rockies on north to Alaska, heavy snow drives the forest giants to lower elevations where the frequency of encounters with dogs, vehicles, and humans rises. Just as moose lose fear of wolves and bears when predation is relaxed, they respond similarly when not harassed by humans. Although attacks occur in areas where animals are both harvested and habituated, once moose become inured to humans, pilo-erection of nape fur, flattened ears, and aggressive charges become more frequent. Moose are more skittish when they first arrive in an unfamiliar setting, but as winter progresses, apprehension turns to tolerance. At some point nutrient-valued shrubbery, plowed driveways, or alfalfa are discovered. Some well-intentioned people offer bread, even frozen apples and carrots. As moose become less wary, what were once trivial, almost imperceptible, threats toward humans turn more serious. Boldness increases. Injuries follow. The extreme is death.

The path from naiveté to danger is bidirectional. Human safety is jeopardized when animals become habituated, but when humans have no fear, animals in the wild are the losers. Travelers lacking knowledge of animal behavior or failing to adopt appropriate precautions die in a process that eventually causes the deaths of animals. Cases are numerous; nascent hikers in grizzly bear country, unwary campers with lions in Kenya's Tsavo, or photographers anxious for a full-framed elephant. Irrespective of our naiveté or blatant stupidity, concerns for human safety trump all. Animals are killed as a result.

"I THINK IT is fear and distrust of the human scent, rather than dislike of human flesh, that prevents the majority of [African] lions from attacking man deliberately," said Charles Astley Mayberry in 1963. Elizabeth Marshall Thomas went further. Based on her experience in the Western Kalahari, which began in the 1950s, Thomas summarized her impressions of how lion culture changed during forty years of interactions with humans.

Initially, lions and !Kung hunters lived in cooperative but independent societies, each pursuing similar game—eland, springbok, gemsbok, and giraffe. Europeans had avoided the harsh Kalahari while concentrating more on African coastal regions, and the lions and indigenous humans lived side-by-side for thousands of years. Thomas recalled a !Kung man telling her, "The lions around here don't harm people. Where lions aren't hunted, they aren't dangerous." It was claimed to be an uneasy truce, with both species paying attention to the other.

With the passing of time, lands that stretched from Botswana's parched mopane forests and desert grasslands to Namibia's Etosha Pan fell into private or government hands. By the 1950s, some were appropriated for wildlife parks. As the hunters and gatherers were disenfranchised from what had been their homes, the long, tenuous relationships they once shared with lions ended. No longer did indigenous hunters clothed in loincloths and armed with bows and arrows cross reddened landscapes of desert and dust. No longer did !Kung compete with lions for food. The bipeds merely slipped silently from memory. Tucked safely away from humans in protected parks, lions presumably lost their apprehension to humans, as all they saw were vehicles. Finally, when humans strolled from the protection of their cars, the emboldened lions did not know how to act. Lions in western Etosha became aggressive to bipeds. They attacked the vehicles of researchers, chased others, and, unlike the !Kung who once walked with lions, the going line was that it was far more dangerous for unarmed biologists to do so. The behavior of lions had changed.

WHEN ANIMAL AND human cultures intersect, animals suffer. Perhaps the human hands-off approach in parks has made lions more dangerous. Lions are smart. They figure out what is dangerous and what is not. They don't know cars are driven by humans. They don't see humans strolling across savannas. They are unwary of all but unfamiliar lion prides and a handful of dangerous elephants. They have nothing to fear. They are on the other side of the Darwinian divide.

chapter 17

Of Fear and Culture

> I had been racked by the weariness of long marches through wind-whipped dunes . . . the fear of raiding parties [kept] us alert and tense. . . . Always our rifles were in our hands and our eyes searching the horizon.
>
> WILFRED THESIGER (1959)

WHILE TRAVERSING DESERT sands in Saudi Arabia's Empty Quarter half a century ago, Thesiger might easily have been ruthlessly hunted by his enemies. To stay alive, the knighted Brit became animalistic, the risk of death keeping him nimble. Like a gazelle or oryx, his first line of protection was vigilance and fleetness. Ingenuity was his second. Living as a modern human, a gun was his third.

Animal defense has existed for eons—concealment, coloration, even the acrid smell of a frightened skunk. So too have protective structures. The armored plates and thickened dermal shields of male Indian rhinos are used to fend off the damage of a competitor's razorlike elongate incisors. Male mountain goats and impala have thickened tissue and skin known as dermal shields where horn penetration by rivals is likely. In other species, protective adornments coat both sexes. Pangolins have hardened scales; porcupines and echidnas sport barbed quills or painful spines.

Armament and defensive behavior have shared similar pathways, both products of evolutionary arms races. Just as individuals respond to the danger of a member of the same species, individuals also respond to species beyond their genetic sphere. Predators, by removing less effectively equipped individuals, have honed the protective traits of survivors.

Like animals, defense in humans has relied on parallel actions—concealment, alertness, and flight. Life in groups is a more recent social innovation. To maximize survival we embrace group living as our ancestors undoubtedly did. *Homo sapiens* did not evolve as armed pursuers of big game, despite our prowess as hunters for the last forty or fifty thousand years or more. To the contrary; our behavior has been molded by half a million years as prey. Today's decisions to avoid danger are grounded in choices modified by internal and external biological forces.

We elect not to walk down dangerous alleys. We avoid perilous parts of cities. We dine in safe areas. In crime-ridden neighborhoods in Johannesburg or Luanda, Los Angeles or New York, the threat of possible attack restricts our movements. If well nourished, we might opt not to dine out. Prudence trumps satiation. Like a well-fed bear, we hunker down and stay put.

If wretched with hunger or teetering like a hummingbird on the precipice of an energy meltdown, we would do things differently. To find food, we would pass quickly through places perceived as dangerous. If food were only in those zones, we might buffer ourselves with additional bodies and randomize movement. Such types of assemblages have a social and ecological context: ungulates herd, birds flock, fish school. Human actions are more animalistic than we credit. Steeped in maneuvers of ancient defense, Wilfred Theisger and his gazelles endured Arabian sands by behaving similarly, moving under the cover of darkness, being unpredictable, and coursing in groups.

DEGREES OF THREAT guide both animals and humans. The former modifies its use of habitats. We do likewise. In safe areas, vigilance is relaxed, doors go unlocked. Our security and valuables receive but a passing thought. Parental vigilance is muted. As danger increases, windows are barred, homes gated, razor wire strewn. Knives and guns become normal defensive armament.

With continued risk, offspring, friends, even strangers adopt many of the same traits, and these are then passed to subsequent generations. Lighted yards and eternal vigilance easily prove adaptive. Others, the simple outgrowth of past superstition, are wrought by groundless fears seamed in tradition. In past times, both in South Africa and the United States, those suspected of being evil because of an alleged affinity to witchcraft were killed. The practitioners of such crimes often justified their actions by reliance on witchcraft as defense. However absurd, such draconian behavior reflects preemptive action to safeguard against primal fears.

Still, the threat of injury permits anxiety and trepidation to run deep. Among pubescent Saharan women, the prospect of genital mutilation understandably incites apprehension. From Somalia and Djibouti to Mali and Guinea-Bissau some one hundred million have already experienced this unimaginable centuries-old tradition.

Since the twelfth century, a different ritual has brought terror to the marriage-aged females of Kyrgyzstan and has led to the adoption of defensive behavior to thwart the perpetrators. In a ritual known as *ala kachuu*, males in this former Soviet republic kidnap, harbor, and force young women to submit. They then marry. Ainur Tairova's torturous abduction in 2001 was no exception. Despite screaming, fighting, and then refusing to eat, she consented to marriage only after her parents arrived at the kidnapper's home and convinced her to acquiesce. She now claims her husband is a good man.

The tactics used by college-aged Kyrgyz females like Ainur to avoid predaceous males share common elements. Females often remain at home, deriving protection from their parents rather than moving about on their own. Others regularly coalesce in groups, and some offer illicit claims of lost virginity. To deceive men into believing they are no longer available, some advertise with head scarves known as *jooluks* or wedding bands. *Ala kachuu,* while illegal, may still typify about thirty percent of the Kyrgyz nuptials.

Fear has its costs whether rational or irrational, whether due to thoughts of kidnapping, violence, or a repressive government, whether human or animal. Options become reduced; for example, food inevitably may decline if one restricts movements because of an increased risk of detection. Where groups are joined, competition may intensify for limited supplies. If we become more vocal, we run the risk of being singled out. Fear may sacrifice freedoms.

There are also benefits. Mechanisms are put in place to minimize dangerous encounters. Vigilance and caution can increase longevity and dampen the opportunity for deadly attacks. The key is balance, disentangling the justified from the unfounded.

FEAR SPELLBINDS. The media bombards us—war and tragedy, violence and victim—daily. Biology's grip is unrelenting. Beyond food and family, air and water, fear is visceral, a feeling shared and similarly exploited by both humans and by animals.

In the extreme, individuals play on apprehensions of death for personal gain. To garner votes, presidential candidates often claim that a lack of armament and aggressive deterrence will lead America to economic vulnerability and invasion. Perhaps correct, perhaps not, but the truth matters little; scare tactics and bravado work almost as well in animals as in humans. Rather than the allure of an election victory, British sparrows set desires on less lofty goals—enhanced access to food. Like humans, subordinate birds also adopt vocal tactics. Theirs occur at feeders where they feign alarm of predators, causing dominants to flee. In doing so, they too play on fear to get seeds.

Birds and humans obviously differ, yet behavioral similarities exist. Exploitation of fear is just one. The cultural transmission of dialects is another. In birds, regional differences in song develop. For humans there is also regional variation in the same language. Other parallels persist—the gradation between aggressive defense and offensive deterrence.

Protective armor in animals has been honed by past pressures much like human defensive tactics. Whereas we once used shields and wielded clubs, individual armament has grown increasingly deadly. As early as 1451, Ma Huan, a Chinese Muslim described arms in Java, noting that boys and old men carry knives and, "If a man touches their head with his hand, or if there is a misunderstanding about money . . . they at once pull out these knives and stab [each other]." These days, it is high-velocity, steel-penetrating bullets that have replaced simple weapons, and avoidance escalates to threat and bluff, even to preemptive attack. Military tacticians and evolutionary biologists embrace game theory as a means of understanding the inevitability of arms races and why mixed and unpredictable strategies are requisite to effective defense.

Abject fear has yielded an interesting extreme in legislation passed in Florida. Known as the "stand your ground" bill, no longer must people make a reasonable attempt to avoid confrontation. If one is a licensed gun holder over the age of 21, protection is available without even feigning escape. "Shoot first" can be the mantra, a behavior that extends far beyond individuals.

The term "preventive war" emerged formally in 1947, two years after Hiroshima. In 1948, Edwin Hopkins produced a song called "What Are We Waiting For," which included the lyrics: "Oh, we've got the bombs to do the job. Why let the despots thrive . . . blast them to kingdom come. Let not the savage gang survive."

Fear leads to overt behavior. While the defense tactics of animals include flight, concealment, or group formation, among humans defense has grown more extreme and now includes preemptive military attack.

I BEGAN THIS book by posing three central questions: (1) Can naive prey avoid extinction when they encounter reintroduced carnivores? (2) To what extent is fear transmitted culturally? (3) How can an understanding of current behavior help unravel the ambiguity of past extinctions and still contribute to future conservation? I also put forth a desire to share in a pursuit that is far from pure, the amalgamation of science with wildlife conservation.

The journey has taken more than 15 years. It whetted my appetite to learn and introduced me to fascinating animals and gallant people. There have been the poor and the rich who continue to give selflessly to preserve our natural heritage. There have been the crazed and the brilliant. There has been acrimony, ego, and uncertainty. Conservation has a human face. It parallels life—its passions, betrayals, successes, and failures.

Answers to the above three questions have come in spurts and only with partial resolution. Gains have been made and lessons learned. They vary from individuals to populations, from North America to Africa, and by species—past to present.

FIRST, THE VULNERABILITY of the naive to localized extinction. We and the world now know that both elk and moose that were once wolf-free have not gone extinct after lobos returned. Both prey species showed initial and extreme naiveté, failing to note the sounds of their ancient enemy. Within a single generation, they became savoir faire. Moose from Wyoming had grown almost as nonchalant to bears as they had to wolves, generally disregarding the odors of grizzlies—at least to a far greater extent than did than their Alaskan cousins. The behavioral naiveté that led to initially high calf losses in both species was followed by learning within a single generation.

Whether unknowing populations of other prey, such as caribou on the Arctic islands of Svalbard or Greenland, can learn about predators as quickly is unknown. Populations on mainland Alaska, on the other hand, remain savvy. Bison differ, remaining indifferent to wolves. Those having neither seen, heard, nor smelled predators for 75 to 100 years behaved like those living with wolves in northern Canada.

For species that join the food chain and live beyond Arctic shores and North America's mountains and prairies, knowledge about extinct but once-familiar predators seems to spread rapidly. Norwegian moose become more difficult quarry for brown bears once coexistence reoccurs. European red deer in Patagonia respond to cougars like elk from Montana. African antelope become hypervigilant after experiencing once-extirpated cheetahs and lions. The return of familiar predators reinstills fear. Perhaps all that is needed is a semi-familiar form—one with four legs—from one's distant past. How distant is distant is more than semantic. Deep time may or may not bury the secrets of survival.

During the four million years that the modern pronghorn of the genus *Antliocapra* have been around, some 99.8 percent of their time has overlapped that of large and fleet carnivores capable of eating them. The mere ten thousand years cessation in this coassociation has been insufficient for pronghorn to change from their hardened survival strategies honed by evolution. In Madagascar, both sifaka and ring-tailed lemurs respond with fright to small raptors that are incapable of killing them although a large eagle that likely preyed upon these lemurs is now extinct. Are the reactions we see today the by-products of the past? How long is sufficient for responses to be dropped from a species repertoire if they have no value?

Species on islands and isolated for thousands of generations from carnivores are less likely to recognize the threat of predation than their counterparts on the mainland. When unfamiliar species invade islands or the mainland and are so different, prey are often incapable of avoiding them. Australia's celebrated kangaroos and wallabies, as well as their more cryptic marsupial cousins, are a case in point; many are now experiencing serious challenges to their survival. Although we do not know if Tasmanian tigers were ever distinguishable from other carnivores, alien or native, what is clear is that Australia's fauna has fared poorly with dingoes and now more so with introduced foxes and cats.

For most species however, the interplay among naiveté, predator awareness, and population persistence remains unresolved, because there is a dearth of knowledge. Questions regarding how quickly individuals learn about predators, whether the acquisition of such information lead to gains in fitness, and what sorts of variation are inherent within populations have never been explored. Additionally, we know little about most carnivores themselves, particularly about how individuals respond to their major predators—us. If one is interested in understanding how animals learn and/or how to sustain wild populations, these sorts of questions need to be answered. Surprisingly, we still understand very little about behavioral variation within and between populations, whether individuals respond differently to us than to other carnivores, and the extent to which external forces change these relationships.

The acquisition of such knowledge is of more than passing academic interest; it affects how species are managed, restored, and survive. Naive animals will always be challenged as people decide to abet their persistence or speed their demise. Elephants moving from the protection of Kruger National Park northward into Mozambique are unlawfully shot as they stand fearless. The grizzly bears of southwestern Alaska were soon destined for a similar fate with administrative blessings. Those within a thirty-square-mile area of the Douglas River and bounded by the protected reserves of the Katmai and McNeil rivers have grown unafraid of humans during a twenty-year hunting hiatus. Tourists now visit to experience and photograph bears. Recent legislation by the Alaska Department of Fish and Game scheduled an authorized hunt of these bears that have no fear of humans. Whether a slaughter was going to ensue and whether the level of shooting would trigger a shier existence for the surviving bruins was uncertain. Fortunately, it remains unknown, because public outcry and outside pressures forced the Alaskan board to reverse their decision in 2007. The hunt was cancelled; the bears remain unwary.

THE SECOND QUESTION concerned the extent to which fear, or more appropriately, the knowledge of the danger of predation, is culturally transmitted. The quandary is not in defining nor understanding culture. It is in the mobilization of appropriate data. Culture, after all, is nothing more than the spread of knowledge and its use between members of the same pair, group, or troop over time. Geographical variation is a clear characteristic. Ideologically, culture is the description of variation and its social transmission.

Whether Kyrgyz or cowboy, elephant or moose, any discussion of culture hinges on the presumption of a societal library that houses a reservoir of prior knowledge. For Kyrgyz females, it may be passing perceived dangers of marauding males associated with *ala kachuu* to their daughters. For a select minority of wranglers, it may be relaying nightmares of surplus killing by wolves to their youngest kin.

In advanced social animals like elephants, experience and learning transpire over a lifetime. Older matriarch females retain the most savvy when it comes to distinguishing strangers, and they do so far more readily than family groups led by younger females. Since old females also have the largest tusks, they are preferred by those trading in ivory. When the matriarchs die at the hands of a poacher, knowledge of the enemy is likely dampened or erased. Even when orphaned calves survive, their mother's wisdom is no longer available.

Insights also come from asocial species. Wariness has been lost from moose on Kalgin Island and in the Teton region. On the Alaskan mainland, it remains. Fear has been rekindled as some Teton mothers now deem four-legged mammals with big paws, wide heads, and dangerous canines as more than an anomalous threat.

Beyond immediate responses, knowledge has also been passed from mothers to young. This type of transmission is what psychologists and anthropologists refer to as vertical, from one generation to the next, to distinguish it from transfers that involve members of the same cohort or same generation. An understanding of how culture is likely to operate in moose has come about, because some populations are hunted by humans, others by carnivores. Some populations remain free of predators, while the status of others has changed. As long as this sort of variation spans space and time and individual behavior is known, then issues associated with roots to culture can be examined. What is known for this ungainly forest dweller can be placed into three or four untidy packages.

Mothers modulate the responses of their young to humans. Some adroit mothers figure out whether humans are dangerous. When afraid, their senses heighten and they run. Young adopt similar behaviors and pass their socially acquired knowledge about humans to their subsequent offspring. Among the most interesting clues to this process are those that have come from changing the radio collars of known mothers. If from areas with human hunting, calves remained afraid. They fled from us, just as their mothers did. They also fled when their mothers lay immobile during our handling. By contrast, the offspring of nonhunted but restrained mothers remained at their side, even as we approached. Three-hundred-pound calves flattened their ears and raised their hackles. These youngsters showed aggressive flair in blind defense of their helpless mothers. They didn't run.

Independent yearlings retain behaviors similar to their mothers. In other words, when year-old females live on their own, their actions are reflective of their mothers' knowledge of predators. In Alaska, solitary yearlings feared the sounds and smells of wolves or bears. In the Tetons, they did not. Beyond these antipredator behaviors, mothers exerted other effects on the distribution of their offspring.

Offspring adopt their mother's migratory habits. Each year, about 65 percent of the Teton moose moved distances more than twenty miles, the total reflecting migration each spring and autumn. The distance exceeded sixty miles for some. Daughters following their mothers as calves adopted the same behavior when independent of maternal influence. They took on the same migratory habits. Either they moved long distances or remained relatively sedentary.

Moose mothers played other roles, infusing knowledge about foods and habitats into the repertoire of their offspring. In Michigan, young learned what to eat by watching their mothers. In Wyoming, young females developed preference for the habitats of their mothers, ranging from willows and spruce forest to aspen and sage. The implication—while anything but experimental—is that mothers are social architects, shaping the behavior, migrations, and patterns of settlement of the next generation.

In her book *Mother Nature*, Sarah Blaffer Hrdy summarized a mother's gift: "Through her example and direct teaching, a mother shapes [for her offspring] critical assumptions about how the world works, what there is to eat, who there is to be afraid of, who is likely to be well-disposed, and so forth—myriad units of culturally transmitted information."

From whales to elephants, lions to llamas, and monkeys to meerkats, a rich fabric of local knowledge is transmitted within social units. Some concerns danger, some concerns food. We know a little about a few organisms. We know nothing about many. The field is wide open.

THE FINAL QUESTION focused on how current behavior might help understand past extinctions and still contribute to future conservation. In the prior chapter, I summarized the mixed evidence on naiveté as a cause of prey vulnerability to colonizing humans. The presumption of innocence has never been rigorously tested.

Many animals lack an intrinsic fear of humans, especially those that have not been persecuted. Perhaps their ancestors were fearless. We know with certainty that we have altered the behavior of today's survivors, indeed, their very existence. We continue to do so.

Understanding the effects of ancestral bipeds on the behavior of past prey is exceptionally difficult. Recall that human immigrants to North America who arrived ten to thirteen thousand years ago were an alien—not a native—species. The nuances of learning and the experiences of past non-surviving prey to predators remain totally unknown. Even if we assume extinct prey learned about carnivores as rapidly as do some living species, the historic densities of humans and prey, the past efficiency of human hunters, and the degree of reliance on alternate foods during Paleolithic times are vastly uncertain. Yet, it is precisely these sorts of factors that were likely to have distinguished prolonged persistence from swift demise. Today's world will never know how yesterday's extinct prey viewed aliens or whether behavioral frailties turned vibrant lives to nightmarish endings. That is in our past.

MY FORAYS TO foreign places, visits with colleagues, and face-to-face meetings with indigenous peoples were as humbling as they were disquieting. My entries into the homes of honest and hardworking cowboys were no different.

The conservation field is daunting, just as the behavior of prey and predator is fascinating. The fact that animals differ as individuals, by personalities and by a mix of experience and genetics, is something that we humans rarely consider. Individuals develop repositories of knowledge that accrue over their lifetimes, a thought given scant attention. We are so arrogant, pretending to know all, yet knowing so little about so much.

Still, we possess and can control future options. Where goals are to maintain or restore ecosystems, insights about fear, behavior, and relationships between predators and prey can dictate what has been lost and what can be retained. As species are disappearing even before formal descriptions have been recorded, so too are behaviors, some subtle, others visible spectacles.

Already we know that responses have been decoupled from systems that once harbored carnivores. Where wolves or grizzly bears are gone, ravens no longer elicit responses by elk or moose. With the disappearance of lions and leopards, giraffes are hardly vigilant and even recline. On a broader scale, however, little is known about the loss of behavioral phenomena or how far back in time we must go to mine for other lost interactions, let alone their importance.

In contrast to indirect human effects vis-à-vis predator removal, the interplay between the global wonder of migration and natural selection may be ending at our hand. Where bison attempt to move beyond the boundaries of Yellowstone Park, they are shot, thus terminating all chances for the transmittal of their knowledge. Those not intent on food beyond the border enjoy a reprieve. Consequently, a form of selection, while anything but natural, plays out. Government policies and agricultural agents dictate which behaviors, migratory or sedentary, are sanctioned for retention. The migratory forms—those selected naturally for thousands of generations—are not favored. Sedentary ones are.

One way to capitalize on knowledge of prey and predator is to appreciate the lessons learned—from both successes and failures. The reintroduction of captive wild dogs into the wilds of Namibia's Etosha National Park is valuable, because it did not work. The spotted dogs knew nothing about lions. They did not know how to hunt efficiently, nor were their parents around to facilitate prey capture. They died while unabashedly attempting to appropriate the kills of lions. By contrast, wild Canadian wolves were brought to Yellowstone. There was no need to teach them to hunt, since their major prey in northern Alberta were elk, just as in Yellowstone.

A final insight derives from the prey perspective. Numerous bighorn sheep reintroductions were botched, because scant attention was given to antipredator strategies. Rather than placing them in large holding areas with low vegetation and good visibility, sheep were maintained in naturally vegetated enclosures where they could not see well. Cougars feasted.

A different sort of application stemming from prey and predator inter-

FIGURE 46. Guanacos in a family band near the Atlantic Ocean in Patagonia.

FIGURE 50. Pronghorn in the Greater Yellowstone Ecosystem.

FIGURE 47. Andres Novaro and Kim in La Cheena, Argentina.

FIGURE 48. Displays of introduced wildlife—red deer antlers and a wild boar—adorn a gaucho's abode in Patagonia.

FIGURE 49. Dingoes, like wolves, are pack-living.

FIGURE 51. Preparing to dart a moose from helicopter for collar replacement.

FIGURE 52. Calf pilo-erection in response to our approach of its immobilized mother.

FIGURE 53. A malnourished female despite surfeit of green. She died three days after this photo was taken. Her calf disappeared and was presumed dead.

FIGURE 54. A moose in piloerection confronts grizzly scats (wrapped in toilet paper).

FIGURE 55. A female that died during birth. Note the circular area of vegetation removal by the struggling female. *Top*, teeth of the calf adjacent to the mother.

FIGURE 56. the real moose investigates the fake one (the author).

FIGURE 57. Archie Gawuseb with moose in Wyoming.

FIGURE 58. Rhino males testing each other in Etosha, Namibia.

FIGURE 59. Horns on a live, dehorned black rhino in Namibia.

actions emanates from efforts to restore carnivores. As mentioned earlier, species as different as river otters and lynx, grizzly bears and lions, are being repatriated to areas where they were once extirpated. Species lacking in recent exposure—fish, snowshoe hares, and zebras—will inevitably become targets, perhaps even black bears when making renewed contact with grizzlies. Understanding their responses in these dynamic systems, the changing nature of behavior, possible redistribution, and food webs will ultimately dictate human tolerance for these broad experiments in ecosystem health and conservation.

By contrast, most landscapes will never have larger carnivores restored. The broader challenge will be to understand the consequences of loss. Ironically, the absence of predators will not necessarily erase the "fear factor" from the repertoire of prey. Associated threats already fill the void left by lost carnivores, doing so in ways analogous to those instilled by apprehension of predators. Humans will always loom large. So might feral species, cats, rats, and dogs, along with other alien replacements. Additional angst will stem from range expansions in coyotes, cougars, or something else. In the United States, as elsewhere in the world, the numbers of hikers, mountain bikers, and all-terrain sport vehicles have surged colossally, and these will excise a toll by displacing animals from areas which offer greater security from us bipeds.

Understanding an animals' ethos with respect to humans and changing environments is equally, if not more, important than understanding their relationships with carnivores, given the inverse relationship between proliferating people and diminishing carnivores. The study of behavior is not, nor will it ever be, the road to conservation riches. It can, however, lead to insights that otherwise would be unavailable. Unquestionably, it is also fun to watch animals, and it fuels a public appetite to learn about natural history and conservation.

The study of animal behavior can accomplish much. Land managers and conservationists need to know about habitats and ecosystems, the mosaic that necessitates both food and refuges from danger, and how animals use them. If healthy numbers of carnivores, condors, or chirus are desirable, we need a gentler footprint. Snow machines ascending mountainous retreats where wolverines raise kits, four-wheelers displacing elk, or hang gliders hovering above sensitive aeries are neither the type nor intensity of experiences these species share with native predators. The effect of fear and avoidance will never be understood unless we understand behavior. Where effects are trivial, mediation will be unnecessary. The true difficulty arises when effects are real. It is then that we need also to understand the human milieu and how we transform cultures into more accepting ways to live with animals.

IN A NATION whose population will soon burgeon from three hundred to four hundred million, where lives of excess exceed those of restraint, and where toys for outdoor recreation continually invade increasingly remote areas, recipes for success can come from a single source—us. The issue is about neither partisanship nor political philosophy. It is about land use. It is about wildlife and conservation. It is about people—a world we inherited and a world we will leave behind.

If the public is blind to the essence of wild animals, to the legacies of so many who worked so enduringly to protect them, and to the splendor of animals themselves, we have failed. What is a landscape without the roar of a lion, the howl of a wolf? What does it mean when savannas grow silent? Are the ghosts of bison thundering across prairies to be joined time and again by the ghosts of others?

When elected officials are not moved by an elk bugling in an early morning fog or by the rippling muscle of a hunting grizzly, by the luminescence of a purple butterfly in a gentle sun or by the eerie trumpeting of an elephant, we must work harder. If deafened by the lure of dollars, we must speak louder. When the world has lost its wild retreats, when our hearts no longer quicken for fear of a wild animal, and when all that is left behind are the haunting memories of a distant but more glorious past, we will all be poorer. Our children will never know what their parents squandered. Indifference guarantees inaction. Only action will fashion conservation victories.

"What is life?" asked Crowfoot, a Blackfoot Indian, in 1890. "It is the flash of a firefly in the night. It is the breath of a buffalo in the winter time. It is the little shadow which runs across the grass and loses itself in the sunset."

Epilogue

I look out my office window in the Tetons and see aspens quivering in a gentle breeze. The early morning quiet is broken by the rhythmic pounding of a ruffed grouse hidden in the forest. The air is cool. Slowly, the darkened sky becomes a muted deep blue. A few weeks earlier, a moose with her new reddish calf crossed below our wooden shed. It is July, yet winter's long grip is not too distant. Snow caps the ridgelines of the Big Hole Mountains to the west, the craggy peaks of the Snake River Range to the south, and the meadows just 1,500 feet above our home. To the north are the Centennials, where the Continental Divide separates Montana from Idaho and from where water spills eastward to the Gulf of Mexico or westward to the Pacific. The Bitterroot Mountains are just beyond. The giant caldera of Yellowstone is north.

The meadows and mountains are clean and beautiful. Fox Creek gurgles past. A few more weeks of warmth, and dragonflies will arrive to skim its watery surface. A warbler will sing. A kingfisher will dive. Just below the steep shelf of Death Canyon only ten miles away, a grizzly bear roams. The deep paw print of a cougar in soft mud fills slowly with water. Last month, a band of elk visited before moving to the high country. Life is everywhere. The world seems peaceful, almost easy.

From my office, I think of the past and the future. I was fortunate to have initiated a program in the southern Yellowstone almost three years before wolves arrived and to have been able to continue it. I experienced remarkable opportunities to contrast systems with and without large carnivores not only here but in sites that stretched from Alaska to the Russian Far East and well beyond. I have watched as the fatally tame succumbed to predators. I have watched as individual animals learned to develop fear, just as humans have accepted life with wolves and bears. Ecosystems have changed, some in part because of our human influence—mine included—others because carnivores have been restored; climate change will continue to exacerbate these alterations. Predictions of prey extirpation have not been borne out. The Yellowstone system is now more diverse, more mysterious.

I move further back in time, another twenty years. When I first visited Yellowstone in 1973, grizzly bears were in decline. The possibility of wolf reintroduction was inconceivable. The chief park scientist suggested that I return to California. Then, in Jackson Hole, Wyoming, a policeman escorted a young hitchhiker—me—to the edge of town. My "type" was not welcome. I muse at the irony.

This is now the place I live, work, and cherish. My youthful innocence did not help me then. In the time it took for me to get from there to here, I have had some seasoning but retain my ideological leanings. Things have changed because people demanded it.

I switch on National Public Radio and absorb the morning news, a touch of despair mixed with hope. Just another untamed day in America. Sunlight drenches the valley below me, bringing farmhouses into view. Life is paced slowly. I think of distant cities and commuters fighting traffic on their way to work. I would like to believe that some reflect back on less frenzied times, when the world was more gentle, when space more open, when wildlife dotted their landscapes. I like to believe that they would enjoy and argue for open space even when living far from it. I deceive myself into believing that my rural neighbors and I are the stewards of a land others would love to visit, a place with a sense of wildness, with unfettered vistas, where wildlife and people commingle.

I return to the sounds of a babbling brook, of kids running through fields adorned with flowers. This is a time of chasing and laughter, a time when youth know nothing of adult burdens. I also look forward and savor the sweetness of budding successes. Wolves and grizzly bears now persist in the Greater Yellowstone Ecosystem, even though they barely cling to a foothold further north along the Canadian border. Bald eagles and peregrines are no longer endangered. Wood bison dot the northern forests. Black rhinos have escaped extinction.

In 1971, when I watched my first, and only, wild condor soaring near Los Angeles, I imagined it would be the last I would ever have the chance to see. Now, they are back in the skies. Black-footed ferrets have been returned to prairies. Neither species is guaranteed a future in the wild, but at least they are there. We have made progress. Battles lay ahead. So does hope.

Edens still exist. They may be neither as large nor as plentiful as those past, but they continue to contribute to the richness of our planet and to our souls. They are living testimony to our past, and to the enormity of future challenge.

As a young man, I dreamed of studying wildlife in natural settings. When people dare to imagine, when they are neither afraid to speak nor to act, vision can turn to victory. I've followed humbly in the footsteps of those with similar dreams, of those far more distinguished, of those with stronger conviction. Yet, in the end, it all comes down to one thing—what are we willing to tolerate?

The question is not about wild or captive, animal or human. It is about all of us—living beings together, in one place, on a single planet. If we become too complacent about the essence of species or their living conditions, then our own lives and our own condition will be haunted by what we have lost, and our paths will merely follow in their shadows.

Acknowledgments

My parents, Alvin and Nathalie, always encouraged learning wherever it took me. While searching, I have encountered conservation's unsung heroes, individuals far afield who are respectful of animals while being sensitive to human culture. Among them are Ronny Aanes, Lu Carbyn, Linda Kerley, and Amy Vedder. Along the way, Marc Bekoff and George Fisler opened the door to graduate school and honed my career choices, as did the late Bill Hamilton III and John Eisenberg. Chris Wemmer pointed to projects beyond borders and reinforced the merit of conservation when too many academicians scoffed at applied endeavors. Mark and Delia Owens walked a different path and are proof that individuals can make a difference.

Organizations that generously contributed funds or other support include the Charles Engelhard Foundation, the National Science Foundation, the National Park Service (Grand Teton, Yellowstone, Badlands, Denali, Rocky Mountain), the National Geographic Society, the US Fish and Wildlife Service (National Elk and Tetlin refuges), Wood Buffalo National Park, Wyoming Game and Fish Department, Alaska Department of Fish and Game, the Dansk Polarcenter (Denmark), Greenland Department of Home Rule, the Norsk Polarinstitute (Norway), the Governor's Office of Svalbard (Norway), and Safari Club International. The Liz Claiborne–Art

Ortenberg Foundation has supported conservation since their inception. If more organizations had the commitment of the LCAOF, the world would be a better place for its wildlife and its people.

At the University of Nevada, Pete Brussard, Erica Fleishman, Lew Oring, and Pete Stacey were stalwart colleagues. Rick Sweitzer shared insights about porcupines and predators; Janet Rachlow did similarly with rhinos and rabbits. Once near-model graduate students with a flair for mischief and adventure, Rick and Janet now champion conservation science for their own students.

In Wyoming, the Earth Friends Wildlife Foundation (Rick Flory and Lee Robert), the Jackson Hole Foundation, the Community Foundation for Jackson Hole, the Denver Zoological Foundation, and Beringea South were gracious with support. Barry Reiswig and Bruce Smith (National Elk Refuge) helped with research permits. Derek and Sophie Craighead and Ted Kerasote offered friendship in Kelly, and Mason Reid assisted with helicopter captures. Steve Cain remains a treasure within the National Park Service, engaged in research and facilitating the park mission. He and Bob Schiller smoothed the way for my work in Grand Teton. The Wyoming Game and Fish Department has always been gracious in issuance of permits and needs more of the ilk of Steve Kilpatrick, Joe Bohne, Susan Patla, and Terry Cleveland. Bernie Holze has been an ardent supporter. Brian Miller and Harry Harlow helped in the field.

In Alaska, Charles Schwartz helped with study possibilities, while Ward Testa offered aerial assistance, ideas, data, and kinship. Terry Bowyer is beyond friendship, having shared meals, his home, and insights for too many years to count. Kevin White is companion par excellence and continues to add to Alaska's ability to conserve wildlife. Steve Amstrup introduced me to polar bears. Mark Mastellar offered housing in the Matanuska Valley while George and Georgina Davidson opened their Oshetna "Hilton" to me.

The National Wildlife Research Center (NWRC) in Millville, Utah, and the Anchorage Zoo in Alaska provided scat and urine samples. John Shivik was most gracious in arranging use of facilities at the NWRC for my sabbatical. Mary Connor and Mike Wolfe at Utah State University enhanced this stint. Toni Ruth and John Goodrich kindly provided large cat photos.

In Mongolia, Rich Reading offered knowledge of argali, khulan, and gazelles as he skillfully covered three thousand fenceless kilometers with me. Sandvin Dulamtseren and Zundvin Namshir explained the nuances of life in the Gobi.

From the far eastern Russia, John Goodrich facilitated travel, guided me to field sites, and opened his home and his heart. He arranged aerial and marine reconnaissance, shared in life and death, and continues to inspire.

He also introduced me to Olga Zaumyslova, who helped me appreciate different perspectives of Russian life. Alexander Mezlinincoff showed me ghoral and sea eagles. Dale Miquelle began the Siberian Tiger Project under the prudent direction of Maurice Hornocker, helped with aerial logistics, and even accompanied me to a rock concert in Vladivostock.

Beyond John Goodrich and Dale Miquelle, passion for conservation runs deep at the Wildlife Conservation Society (WCS). Josh Ginsberg, Justina Ray, and Eric Sanderson shared their knowledge of carnivores and their prey. Andres Novaro and Susan Walker welcomed Kim and me into their home in Junin de Los Andes, introduced us to the importance of exotic prey, and brought us to Patagonia.

George Schaller and Alan Rabinowitz continue to motivate just by being George and Alan. Kent Redford has helped me for more than 25 years. Matt Gompper and Jon Beckmann, once academic colleagues who defected to WCS or beyond, represent the next wave of real-world practitioners. Sanjay Pyare spent more than a year with moose and raised the bar on understanding ecological recovery; Jodi Hilty and Craig Groves do similarly. Steve Zack rightly insists on valuing natural history to promote conservation. John Robinson is living proof that roads to conservation exist beyond behavioral ecology.

It was Bill Weber who began WCS's North American Program and regularly admonished: "Tell me, how your work is saving Yellowstone?" He was right to ask; all researchers need reality checks. If more followed Bill's lead, there would be additional wildlife and more conservation. Steve Sanderson's prescience guides all WCS. Other organizations should be as privileged to have such leaders and colleagues.

My academic partners include Fred Allendorf, Marco Festa-Bianchet, Billy Karesh, Brian Miller, Steve Monfort, Rich Reading, Tom Roffe, Mike Soulé, Pete Stacey, Jon Swenson, and Ward Testa. Tom shows great acumen whether darting moose from the air or handling them in water or deep snow. John Byers has always been an excellent sounding board. Tim Caro taught me to appreciate antipredator defenses, while Josh Donlan and Gary Roemer placed Pleistocene rewilding at center stage. Jim Estes explained why marine systems and those on land share similarities. Harry Greene has been a gallant and courageous spokesperson for conservation well before the discipline gained academic acceptance. For years, Doug Smith has reviewed my work even though he may not always agree with my interpretations. More importantly, Doug promotes an open policy of research in an outdoor laboratory called Yellowstone, and, for a lifetime, has promoted wolf conservation. At the University of Montana, Mark Hebblewhite, Charlie Janson, Scott Mills, Mike Mitchell, Dan Pletscher, and so many others have opened their arms and continue to push for science and conservation.

From the naked deserts of coastal Africa, Archie Gawuseb exemplifies passion for life and for conservation even with little to live on. Closer to the Tetons, John and Karin McQuillan offer support and friendship. Tom Segerstrom contributed valuable information on moose. Deb Markert constructed the moose suit, and Lou Parri engaged in valued discussions of animal cognition.

Among the many field assistants who put data before life, and pain before comfort, I offer my greatest gratitude. I especially thank Mike Nordell, Renee Seidler, and Noah Weber for years of labor, hard-earned data, and the fun we have shared. Noah is probably the only person on the planet to have been bitten by both a moose and a chimpanzee. Helping also in Wyoming or Alaska were Lori Bellis, Paul Brown, Matthew Johnson, Mike McMurray, Jennifer Sands, Doug Spaeth, and Nathan Varley. Jerry Lee supplied the aerial photos of bears from Alaska. Steve Kerr and Michele Peacock immobilized moose in 1995.

Kim Berger, Franz Camenzind, Kathy Gasaway, Karin McQuillan, Mason Reid, and Steve Vessey helped with, corrected, and reedited many chapters. Jonathan Cobb initially encouraged this writing project, but it was my remarkable editor at the University of Chicago Press, Christie Henry, who has been a gifted source of support and enthusiasm. She whittled chapters and forced awkward prose into a read. Such talent is a treasure. Mary Gehl is a talented editor who helped in the manuscript's final phases.

Franz Camenzind remains the heroic voice for wildlife. Without his decades of commitment, Jackson Hole would not be the special place that it is. Steve and Lisa Robertson are true altruists in the fight for wildlife, be it in the Pacific, Africa, or their backyard.

Carol Cunningham remains a truly bright spot in an imperfect world. The mother of my only child, she is unruffled by challenge and brings light wherever she goes. Sonja offers other beauty. Beyond the snakes, fish, mice, and rabbits that prosper in our living room, her drawings enhance gentleness and the splendor of life for all.

Far from last is Caroleena—colleague, confidante, and best friend of Guillermo. When there was darkness she brought joy. Where there is silence she brings laughter and optimism.

Readings of Interest and Exploration

Volumes have been written about the distinct themes explored in this book. This section offers two sorts of literature: general books and papers that serve as a backdrop, and specific publications that arose directly from this project.

General Books and Papers

Aanes, R. 2003. Synchrony in Svalbard reindeer population dynamics. I. *Canadian Journal of Zoology* 81:103–10.
Albanov, V. 1917 [2000]. *In the land of white death.* New York: The Modern Library.
Andrews, R. C. 1932. *The new conquest of central Asia.* New York: American Museum of Natural History.
Arseniev, V. K. 1941. *Dersu the trapper, a true account.* New York: E. P. Dutton.
Barlow, C. 2000. *The ghosts of evolution.* New York: Basic Books.
Beckmann, J., and J. Berger. 2003. Rapid ecological and behavioral changes in carnivores: The response of black bears to altered food. *Journal of Zoology* (London) 261:207–12.
———. 2003. Using black bears to test experimentally ideal-free distributions. *Journal of Mammalogy* 84:594–606.
Bekoff, M., ed. 2004. *Encyclopedia of animal behavior.* Westport, CT: Greenwood Press.
Berger, J., and C. Cunningham. 1994. *Bison: Mating and conservation in small populations.* New York: Columbia University Press.
———. 1994. Active intervention and conservation: Africa's pachyderm problem. *Science* 263:1241–42.
Berger, K. M. 2006. Carnivore-livestock conflicts: Effects of subsidized predator

control and economic correlates on the sheep industry. *Conservation Biology* 20:751–61.
Blumstein, D. T., and J. C. Daniel. 2002. Isolation from mammalian predators differentially affects two congeners. *Behavioral Ecology* 13:657–63.
Blumstein, D. T., M. Mari, J. C. Daniel, J. G. Ardron, A. S. Griffin, and C. S. Evans. 2002. Olfactory predator recognition: wallabies may have to learn to be wary. *Animal Conservation* 5:87–93.
Botkin, D. B. 1995. *Our natural history: The lessons of Lewis and Clark*. New York: G. P. Putnam & Sons.
Bowyer, R. T. 2004. Sexual segregation in ruminants: Definitions, hypotheses, and implications for conservation and management. *Journal of Mammalogy* 85:1039–52.
Box, H. O., and K. R. Gibson, eds. 1999. *Mammalian social learning*. Cambridge: Cambridge University Press.
Byers, J. A. 1997. *American pronghorn: Social adaptations and the ghosts of predators past*. Chicago: University of Chicago Press.
Carbyn, L. 2003. *The buffalo wolf: Predators, prey, and the politics of nature*. Washington, DC: Smithsonian Institution Press.
Caro, T. M. 1999. Demography and behaviour of African mammals subject to exploitation. *Biological Conservation* 13:805–14.
Caro, T. M., and M. D. Hauser. 1992. Is there teaching in nonhuman animals? *Quarterly Review of Biology* 67:151–74.
Cavalli-Sforza, L. L. 1986. Cultural evolution. *American Zoologist* 26:845–55.
Cheney, D. L., and R. M. Seyfarth. 1990. *How monkeys see the world: Inside the mind of another species*. Chicago: University of Chicago Press.
Cherry-Garrard, A. 1922. *The worst journey in the world*. London: Constable.
Clark, T. W., A. P. Curlee, S. C. Minta, and P. M. Kareiva, eds. 1999. *Carnivores in ecosystems: The Yellowstone experience*. New Haven, CT: Yale University Press.
Craighead, J. J., J. S. Summer, and J. A. Mitchell. 1995. *The grizzly bears of Yellowstone*. Covello, CA: Island Press.
Cunningham, C., and J. Berger. 1997. *Horn of darkness: Rhinos on the edge*. New York: Oxford University Press.
Curio, E., and E. W. Vieth. 1978. Cultural transmission of enemy recognition: One function of mobbing. *Science* 202:899–901.
Darwin, C. 1859. *On the origin of species*. New York: Collier Books.
Diamond, J. 1997. *Guns, germs, and steel: The fates of human societies*. New York: Norton.
Flannery, T. F. 1995. *The future eaters: An ecological history of the Australasian lands and people*. Port Melbourne: Reed Books Australia.
Foster, S. A., and J. A. Endler, eds. 1999. *Geographical variation in behavior*. New York: Oxford University Press.
Fragasyz, D. M., and S. Perry, eds. 2003. *The biology of traditions: Models and evidence*. Cambridge: Cambridge University Press.
Franzmann, A. W., and C. C. Schwartz, eds. 1997. *Ecology and management of the North American moose*. Washington, DC: Smithsonian Institution Press.
Frison, G. C. 2004. *Survival by hunting: Prehistoric human predators and animal prey*. Berkeley: University of California Press.
Frost, O. W., ed. 1988. *Georg Wilhelm Stellar, journal of a voyage with Bering, 1741–1742*. Palo Alto, CA: Stanford University Press.

Gittleman, J. L., and M. E. Gompper. 2001. The risk of extinction—what you don't know will hurt you. *Science* 291:997–99.
Guthrie, R. D. 1990. *Frozen fauna of the Mammoth Steppe: The story of Blue Babe.* Chicago: University of Chicago Press.
Hedin, S. 1925. *My life as an explorer.* London: Boni and Liveright, Inc.
———. 1931. *Across the Gobi.* New York: Greenwood Press.
Heinrich, B. 1999. *Mind of the raven.* HarperCollins, New York.
Heller, E. 1925. The big game animals of Yellowstone National Park. *Roosevelt Wildlife Bulletin* 2:430.
Hess, K. Jr. 1993. *Rocky times in Rocky Mountain National Park.* Niwot: University of Colorado Press.
Jones, M. E., G. C. Smith, and S. M. Jones. 2004. Is anti-predator behaviour in Tasmanian eastern quolls (*Dasyurus viverrinus*) effective against introduced predators? *Animal Conservation* 7:155–60.
Kerasote, T. 1994. *Bloodties: Nature, culture, and the hunt.* New York: Random House.
Kruuk, H. 2002. *Hunter and hunted: Relationships between carnivores and people.* Cambridge: Cambridge University Press.
Kurten, B. 1976. *The cave bear story: Life and death of a vanished animal.* New York: Columbia University Press.
Lent, P. C. 1999. *Muskoxen and their hunters.* Norman: University of Oklahoma Press.
Lentfer, H., and C. Servid, eds. 2001. *Arctic refuge, a circle of testimony.* Minneapolis, MN: Milkweed Editions.
Martin, P., and R. G. Klein, eds. 1984. *Quaternary extinctions: A prehistoric revolution.* Tucson: University of Arizona Press.
Matthieseen, P. 1999. *Tigers in the snow.* New York: North Point Press.
McComb, K., C. Moss, S. M. Durant, L. Baker, and S. Sayialel. 2001. Matriarchs as repositories of social knowledge in African elephants. *Science* 292:491–94.
Mech, L. D., and L. Boitani, eds. 2003. *Wolves: Behavior, ecology, and conservation.* Chicago: University of Chicago Press.
Mech, L. D., L. G. Adams, T. J. Meier, J. W. Burch, and B. W. Dale. 1998. *The wolves of Denali.* Minneapolis: University of Minnesota Press.
Miquelle, D. G., P. A. Stephens, E. N. Smirnoff, J. M. Goodrich, O. J. Zaumyslova, and A. Myslenkov. 2005. Tigers and wolves in the Russian far east: Competitive exclusion, functional redundancy, and conservation implications. In *Carnivores and biodiversity: Does saving one conserve the other?* ed. J. Ray, K. H. Redford, R. Steneck, and J. Berger,179–208. Covello, CA: Island Press.
Morden, W. J. 1927. *Across Asia's snows and deserts.* New York: G. P. Putnam & Sons.
Murie, A. 1961. *A naturalist in Alaska.* New York: Devin-Adair Company.
Murie, M., and O. Murie. 1966. *Wapiti wilderness,* New York: Knopf.
Nansen, F. 1903 [1999]. *Farthest north.* New York: The Modern Library.
Ohman, A., and S. Mineka. 2003. The malicious serpent: Snakes as a prototypical stimulus for an evolved module of fear. *Current Directions in Psychological Science* 12:5–9.
Owens, M., and D. Owens, 1983. *Cry the Kalahari.* Boston: Houghton Mifflin.
Petersen, D. 2002. Rewilding the hunt: a conversation with Michael Soulé. *Bugle* (Sept–Oct): 60–71.
Rabinowitz, A. 2001. *Beyond the last village: A journey of discovery in Asia's forbidden wilderness.* Washington, DC: Shearwater Press.

Rasmussen, K. 1927. *Across arctic America.* New York: G. P. Putnam & Sons.

Rawicz, S. 1956. *The long walk.* New York: Lyons and Burford, Publishers.

Ray, J., K. H. Redford, R. Steneck, and J. Berger, eds. 2005. *Large carnivores and conservation of biodiversity.* Covello, CA: Island Press.

Reading, R. P., H. Mix, B. Lhagvasuren, and E. Blumer. 1999. Status of wild Bactrian camels and other large ungulates in south-western Mongolia. *Oryx* 33:247–55.

Ridley, M. 2003. *The agile gene: How nature turns on nurture.* London: Harper.

Ripple, W. J., and R. L. Beschta. 2004. Wolves and the ecology of fear: Can predation risk structure ecosystems? *Bioscience* 54:755–66.

Schaller, G. B. 1998. *Wildlife of the Tibetan Steppe.* Chicago: University of Chicago Press.

Shalter, M. D. 1984. Predator-prey behavior and habituation. In *Habituation, sensitization, and behavior*, ed. Harman V. S. Peeke, 349–380. New York: Academic Press.

Short, J., J. E. Kinnear, and A. Robley. 2002. Surplus killing by introduced predators in Australia—evidence for ineffective anti-predator adaptations in naive prey species. *Biological Conservation* 103:283–301.

Smith, D., and G. Ferguson, 2005. *Decade of the wolf: Returning the wild to Yellowstone.* Guildford, CT: The Lyons Press.

Smith, D. W., R. O. Peterson, and D. B. Houston. 2003. Yellowstone after wolves. *Bioscience* 53:330–40.

Sweitzer, R. A., S. H. Jenkins, and J. Berger. 1997. Near-extinction of porcupines by pumas in the Great Basin and consequences of ecological changes. *Conservation Biology* 11:1407–17.

Terborgh, J. et al. 2001. Ecological meltdown in predator-free forest fragments. *Science* 294:1923–1926.

Testa, J. W. 2004. Population dynamics and life history trade-offs of moose (*Alces alces*) in southcentral Alaska. *Ecology* 85:1439–52.

———. 2004. Interaction of top-down and bottom-up life history trade-offs in moose (*Alces alces*). *Ecology* 85:1453–59.

Thesiger, W. 1959. *Arabian Sands.* New York: E. P. Dutton.

Thomas, E. M. 1994. *The tribe of the tiger: Cats and their culture.* New York: Simon and Schuster.

Van Schaik, C. P., and G. R. Pradham. 2003. A model for tool use traditions in primates: Implications for the coevolution of culture and cognition. *Journal of Human Evolution* 44:645–64.

Walker, S., A. Novaro, M. Funes, R. Baldi, C. Chehebar, E. Ramilo, J. Ayesa, D. Bran, and A. Vila. 2005. Re-wilding Patagonia. *Wild Earth* 14:36–41.

Wallace, A. R. 1881. *Island life.* New York: Harper and Bros.

Weber, B. 2004. The arrogance of America's designer ark. *Conservation Biology* 18:1–3.

Whitehead, H. 2003. *Sperm whales: Social evolution in the ocean.* Chicago: University of Chicago Press.

Wrangham, R. W., W. C. McGrew, F. B. M. De Waal, and P. Heltne, eds.. 1996. *Chimpanzee cultures.* Cambridge, MA: Harvard University Press.

Publications Arising from *The Better to Eat You With*

Berger, J. 1998. Future prey: Some consequences of losing and restoring large carnivores. In *Behavioral ecology and conservation biology*, ed. T. Caro, 80–100. Oxford: Oxford University Press.

———. 1999. Anthropogenic extinction of top carnivores and interspecific animal behaviour: Implications of the rapid decoupling of a web involving wolves, bears, moose and ravens. *Proceedings of the Royal Society* 266:2261–67.
———. 2002. Wolves, landscapes, and the ecological recovery of Yellowstone. *Wild Earth* 12:32–37.
———. 2003. Is it acceptable to let a species go extinct in a national park? *Conservation Biology* 17:1451–54
———. 2003. Through the eyes of prey: How the extinction and conservation of North America's large carnivores alter prey systems and biodiversity. In *Animal behavior and wildlife conservation*, ed. M. Festa-Bianchet and M. Apollonio, 133–156. Covello, CA: Island Press.
———. 2004. The longest mile: How to sustain long distance migration in mammals. *Conservation Biology* 18:320–32.
———. 2005. Hunting by carnivores and by humans: Is functional redundancy possible and who really cares? In *Large carnivores and conservation of biodiversity*, ed. J. Ray, K. H. Redford, R. Steneck, and J. Berger, 316–41. Covello, CA: Island Press.
———. 2007. Carnivore repatriation and holarctic prey: Narrowing the deficit in ecological effectiveness. *Conservation Biology* 21:1105–16.
———. 2007. Fear, human shields, and the re-distribution of prey and predators in protected areas. *Biology Letters* 3:621–24.
———. 2008. Undetected species losses, food webs, and ecological baselines: A cautionary tale from Yellowstone. *Oryx* 42:139–43.
Berger, J., S. L. Cain, and K. Murray Berger. 2006. Connecting the dots: An invariant migration corridor links the Holocene to the present. *Biology Letters* 2:528–31.
Berger, J., S. Dulamtseren, S. Cain, D. Enkkhbileg, P. Lichtman, Z. Namshir, G. Wingard, and R. Reading. 2001. Back-casting sociality in extinct species: New perspectives using mass death assemblages and sex ratios. *Proceedings of the Royal Society*. 268: 131–39.
Berger, J., and M. E. Gompper. 1999. Sex ratios in extant ungulates: Products of contemporary predation or past life histories? *Journal of Mammalogy* 80:1084–113.
Berger, J., A. Hoylman, and W. A. Weber. 2001. Perturbation of vast ecosystems in the absence of adequate science: Alaska' Arctic Refuge. *Conservation Biology* 15:539–41.
Berger, J., and D. Smith. 2005. Restoring functionality in Yellowstone with recovering carnivores: Gains and uncertainties. In *Large carnivores and conservation of biodiversity*, ed. J. Ray, K. H. Redford, R. Steneck, and J. Berger, 100–9.Covello, CA: Island Press.
Berger, J., P. B. Stacey, M. L. Johnson, and L. Bellis. 2001. A mammalian predator-prey imbalance: Grizzly bear and wolf extinction affects avian neotropical migrants. *Ecological Applications* 11:947–60.
Berger, J., J. E. Swenson, and I. Lill-Persson. 2001. Re-colonizing carnivores and naive prey; Conservation lessons from Pleistocene extinctions. *Science* 291:1036–39.
Berger, J., J. W. Testa, T. Roffe, and S. L. Montfort. 1999. Conservation endocrinology: A noninvasive tool to understand relationships between carnivore colonization and ecological carrying capacity. *Conservation Biology* 13:980–89.
Berger, K. M., and M. M. Conner. 2008. Re-colonizing wolves and meso-predator suppression of coyotes: Impacts on pronghorn population dynamics. *Ecological Applications* (forthcoming).

Berger, K. M., and E. M. Geese. 2007. Does interference competition with wolves limit the distribution and abundance of coyotes? *Journal of Animal Ecology* 76:1075–85.

Berger, K. M., E. M. Geese, and J. Berger. 2008. Indirect effects and traditional trophic cascades: A test involving wolves, coyotes, and pronghorn. *Ecology* 89:818–28.

Donlan, C. J. et al. 2005. Re-wildling North America. *Nature* 436:913–14.

Donlan, C. J. et al. 2006. Pleistocene rewilding: An optimistic agenda for twenty-first century conservation. *American Naturalist* 168:660–81.

Pyare, S. and J. Berger. 2003. Beyond demography and de-listing: Ecological recovery for Yellowstone's grizzly bears and wolves. *Biological Conservation* 113:63–73.

Pyare, S., S. L. Cain, D. Moody, C. Schwartz, and J. Berger. 2004. Grizzly bears in the Yellowstone ecosystem: Loss and re-colonization rates during a century of change. *Animal Conservation* 7:71–78

Ray, J. C., K. H. Redford, J. Berger, and R. Steneck. 2005. Is large carnivore conservation equivalent to biodiversity conservation, and how can we achieve both? In *Large carnivores and conservation of biodiversity*, ed. J. Ray, K. H. Redford, R. Steneck, and J. Berger, 400–28. Covello, CA: Island Press.

Soulé, M. E., J. A. Estes, J. Berger, and C. Martinez del Rio. 2003. Ecological effectiveness: Conservation goals for interactive species. *Conservation Biology* 17:1238–50.

White, K., and J. Berger. 2001. Anti-predator strategies of Alaskan moose: Are maternal trade-offs influenced by offspring activity? *Canadian Journal of Zoology* 79:2055–62.

———. 2001. Behavioral and ecological effects of differential predation pressure on moose in Alaska. *Journal of Mammalogy* 82:422–29.

Index